Lecture Notes in Management and Industrial Engineering

Series Editor

Adolfo López-Paredes, INSISOC, University of Valladolid, Valladolid, Spain

This book series provides a means for the dissemination of current theoretical and applied research in the areas of Industrial Engineering and Engineering Management. The latest methodological and computational advances that can be widely applied by both researchers and practitioners to solve new and classical problems in industries and organizations contribute to a growing source of publications written for and by our readership.

The aim of this book series is to facilitate the dissemination of current research in the following topics:

- Strategy and Entrepreneurship
- Operations Research, Modelling and Simulation
- Logistics, Production and Information Systems
- Quality Management
- Product Management
- Sustainability and Ecoefficiency
- Industrial Marketing and Consumer Behavior
- Knowledge and Project Management
- Risk Management
- Service Systems
- Healthcare Management
- Human Factors and Ergonomics
- Emergencies and Disaster Management
- Education

More information about this series at http://www.springer.com/series/11786

Roxana Cuevas · Constanta-Nicoleta Bodea ·
Pablo Torres-Lima
Editors

Research on Project, Programme and Portfolio Management

Integrating Sustainability into Project Management

Springer

Editors
Roxana Cuevas
Department of Sociology
Universidad Autónoma Agraria
Antonio Narro
Saltillo, Coahuila, Mexico

Pablo Torres-Lima
Agricultural and Animal Production
Universidad Autónoma Metropolitana
Mexico City, Distrito Federal, Mexico

Constanta-Nicoleta Bodea
Economic Informatics and Cybernetics
Bucharest University of Economic Studies
Bucharest, Romania

ISSN 2198-0772 ISSN 2198-0780 (electronic)
Lecture Notes in Management and Industrial Engineering
ISBN 978-3-030-60138-6 ISBN 978-3-030-60139-3 (eBook)
https://doi.org/10.1007/978-3-030-60139-3

This Springer imprint is published by the registered company Springer Nature Switzerland AG
The registered company address is: Gewerbestrasse 11, 6330 Cham, Switzerland

Foreword

Limited understanding of human-environment systems prevented business to consider sustainability as one of its long-term goal and responsibility. In our days, the need for achieving sustainability in all human activities is fully acknowledged. For achieving this, the collaboration between research and education institutions, on the one hand, and management and policy bodies, on the other, is critical.

It is considered that systematic research on sustainability started in 2001, with the *Declaration on Earth System science* that was launched during the World Congress 'Challenges for a Changing Earth 2001'. Sustainability science is based on the concepts of sustainability and sustainable development and the methods of measuring sustainability, in order to substantiate policies and governance in this area. Different research programmes in sustainability science were defined, mainly by universities, having as a general objective to assess and minimize the negative consequences of human activities on natural and social systems.

Recently, researchers have linked sustainability with project management, arguing the need for harmonization of environmental and economical interests in projects, the short-term and long-term project orientation, the local and global orientation in projects and the promotion of values, ethics, transparency and accountability in project management. Incorporating sustainability in project management also implies stakeholder participation, risks reduction and eliminate waste.

The book includes recent research on sustainability in project management, presented at the 31st IPMA World Congress held in the Yucatan International Congress Centre, Merida, Mexico from 30 September to 2 October, 2019. International Project Management Association (IPMA) is a federation of 73 member associations, which are offering know-how, products and services to the benefit of individuals, projects and organizations across public, private and community sectors.

Research selected to be included in the book cover different topics of sustainability in project management, such as project excellence by incorporating sustainability, sustainability and project governance, factors affecting the integration of sustainability, incorporating sustainability in projects for smart cities and smart rural, developing competencies for sustainable project management. The book includes 21 chapters, written by authors from 14 countries from four continents. This high geographical coverage assures a high diversity of research contexts and findings.

The book sends an invitation for conducting further research to allow sustainability to be achieved in and through projects. Enjoy reading it!

Professor Yan Xue, Ph.D.
Peking University
Beijing, China

Contents

Contributors

Hans Bakker Delft University of Technology, Delft, The Netherlands

María José Bastante-Ceca Universitat Politècnica de València, Valencia, Spain

Bjarnveig Birta Bjarnadottir School of Business, University of Iceland, Reykjavík, Iceland

Constanta-Nicoleta Bodea Bucharest University of Economic Studies, Bucharest, Romania

Marian G. C. Bosch-Rekveldt Faculty of Civil Engineering and Geosciences, Delft University of Technology, Delft, The Netherlands

Salvador F. Capuz-Rizo Universitat Politècnica de València, Valencia, Spain

Roxana Cuevas Department of Rural Sociology, Antonio Narro Agrary Autonomous University, Saltillo, Coahuila, Mexico

Maria-Iuliana Dascalu University POLITEHNICA of Bucharest, Bucharest, Romania

Jorge Flores Bioagroprojects Biotech PPM, C/Bisbal s/n, Villareal, CS, Spain

Thordur Vikingur Fridgeirsson School of Engineering, Reykjavik University, Reykjavik, Iceland

Amir Ghanbaripour Bond University, Robina, QLD, Australia

Agnes Holm Gunnarsdottir Reykjavik University, Reykjavik, Iceland

Alexandru Hang University POLITEHNICA of Bucharest, Bucharest, Romania

Marcel Hertogh Delft University of Technology, Delft, The Netherlands

Marcel J. C. M. Hertogh Faculty of Civil Engineering and Geosciences, Delft University of Technology, Delft, The Netherlands

Helgi Thor Ingason School of Engineering, Reykjavik University, Reykjavik, Iceland

Afshin Jalali Sohi Delft University of Technology, Delft, The Netherlands

Haukur Ingi Jonasson Reykjavik University, Reykjavik, Iceland

Asa Bjork Jonsdottir Reykjavik University, Reykjavik, Iceland

Milica Jovanović Faculty of Organizational Sciences, University of Belgrade, Belgrade, Serbia

Philipp Klausing Institute of Production Systems and Logistics, Leibniz University Hannover, Garbsen, Germany

Hans Knoepfel Rosenthaler + Partner AG, Zurich, Zurich, Switzerland

Kari Kristinsson School of Business, University of Iceland, Reykjavík, Iceland

Bara Hlin Kristjansdottir School of Engineering, Reykjavik University, Reykjavik, Iceland

Craig Langston Bond University, Robina, QLD, Australia

Jaehyun Lee Department of Applied Engineering, Graduate School of Engineering Practice, Seoul National University, Seoul, Republic of Korea

Unho Lee Institute of Engineering Research, Seoul National University, Seoul, Republic of Korea

Vanesa G. Lo Iacono Ferreira Universitat Politècnica de València, Valencia, Spain;
Universitat Politècnica de València, Alcoy, Alicante, Spain

Adán López Miranda Tecnológico de Monterrey, School of Engineering, Monterrey, N.L., Mexico

Jesus Martinez Almela Bioagroprojects Biotech PPM, C/Bisbal s/n, Villareal, CS, Spain;
IPMA Mexico, Iztacalco, Ciudad de Mexico, Mexico

Sandra Matuhina Ekspertus Management Consulting, Zagreb, Croatia

Inga Minelgaite School of Business, University of Iceland, Reykjavík, Iceland

Maedeh Molaei Faculty of Civil Engineering and Geosciences, Delft University of Technology, Delft, The Netherlands

Maja-Marija Nahod Ministry of Construction and Physical Planning, Zagreb, Croatia

Maryam R. Nezami Delft University of Technology, Delft, The Netherlands

Lars Nielsen Institute of Production Systems and Logistics, Leibniz University Hannover, Garbsen, Germany

Steven A. Nijhuis Utrecht University of Applied Science, Utrecht, The Netherlands

Peter Nyhuis Institute of Production Systems and Logistics, Leibniz University Hannover, Garbsen, Germany

Vladimir Obradović Faculty of Organizational Sciences, University of Belgrade, Belgrade, Serbia

Teri Vivienne Okoro TOCA, London, UK

Ángela Paneque de la Torre Universitat Politècnica de València, Valencia, Spain

Changwoo Park Department of Applied Engineering, Graduate School of Engineering Practice, Seoul National University, Seoul, Republic of Korea

Alan Patching Bond University, Robina, QLD, Australia

Jorge J. Peart Mijangos Bioagroprojects Biotech PPM, C/Bisbal s/n, Villareal, CS, Spain

Mladen Radujković Alma Mater Europaea ECM, Maribor, Slovenia/Project, Program Portfolio Expert, Zagreb, Croatia

Amin Saidoun Alma Mater Europaea, Maribor, Slovenia;
Grenoble, France

Achmad Jaka Santos Department of Civil Engineering, University of Pelita Harapan, Karawaci, Tangerang, Banten, Indonesia;
Postgraduate School, Djuanda University, Bogor, Jawa Barat, Indonesia

Lukas Beladi Sihombing Department of Civil Engineering, University of Pelita Harapan, Karawaci, Tangerang, Banten, Indonesia

Gregory J. Skulmoski Bond University, Robina, QLD, Australia

Robin Tamak Faculty of Civil Engineering and Geosciences, Delft University of Technology, Delft, The Netherlands

Marija Todorović Faculty of Organizational Sciences, University of Belgrade, Belgrade, Serbia

Pablo Torres-Lima Department of Agricultural and Animal Production, Metropolitan Autonomous University, Mexico City, Mexico

Reinhard Wagner Tiba Managementberatung GmbH, Munich, Germany

Andreas Wibowo Department of Civil Engineering, University of Pelita Harapan, Karawaci, Tangerang, Banten, Indonesia;
Agency for Research and Development, Ministry of Public Works and Housing, Bandung, Indonesia

Eunsang Yoon Program Management Consulting Company of Sejong Administrative City Construction (HKCMC), Government Complex Sejong, Sejong-si, Republic of Korea

Chapter 1
Introduction

Roxana Cuevas, Constanta-Nicoleta Bodea, and Pablo Torres-Lima

Abstract This brief introduction provides an overview of the book structure and content. The chapter discusses the concepts, research methods and results which the contributors present in their chapters. The diversity of addressed topics, the wide geographical and sector coverage lead to a wide range of book audiences. The book is useful to researchers, practitioners, educators and trainers in project management domain as well to postgraduate students, in general for all those interested to better understand how the sustainability can be integrated into project management.

Keywords Project management · Sustainability · Sustainable business · Social responsibility · Project excellence · Projectification

Climate change, threats to biodiversity and accelerated exhaustion of non-renewable resources have made researchers and practitioners more involved in revising economic growth models and practices, and in promoting sustainability into theory and practice. The research literature on sustainable development has expanded and more and more companies have started to include sustainability principles into their strategies and to report their regular actions/practices related to sustainable business.

Nowadays, one of the important trends of societies worldwide is projectification. Several studies reveal that around 40% of economic activities are performed as projects and programmes (www.ipma.world). For this reason, the implementation

---

R. Cuevas
Department of Rural Sociology, Antonio Narro Agrary Autonomous University, 14 Chalcatzingo, San Alberto, 25204 Saltillo, Coahuila, Mexico
e-mail: roxanacuevas2000@yahoo.com

C.-N. Bodea (✉)
Bucharest University of Economic Studies, 6 Piata Romana, 1st district, 010374 Bucharest, Romania
e-mail: bodea@ase.ro

P. Torres-Lima
Department of Agricultural and Animal Production, Metropolitan Autonomous University, 23-7 Algodonales, 14300 Mexico City, Mexico
e-mail: ptorres@correo.xoc.uam.mx

R. Cuevas et al. (eds.), *Research on Project, Programme and Portfolio Management*, Lecture Notes in Management and Industrial Engineering,
https://doi.org/10.1007/978-3-030-60139-3_1

of sustainability into project, programme and portfolio management is considered as being critical for assuring sustainable development.

Sustainable project, programme and portfolio management aims to take into consideration the environmental, economic and social aspects of programmes and portfolios in order to achieve sustainable results. Many project, programme and portfolio management professionals took the responsibility for integrating sustainability by developing sustainable objectives and management plans and by translating them into practice. Professional associations, such as the International Project Management Association (IPMA) are actively involved in promoting sustainability in project, programme and portfolio management disciplines by financing research, developing standards and organizing events as platforms for professional knowledge sharing.

The book includes the extended version of selected papers presented at the 31st IPMA World Congress held in the Yucatan International Congress Centre, Merida, Mexico from September 30 to October 2, 2019. The main topic of the IPMA World Congress was "Integrating Sustainability into Project Management". These selected papers address relevant topics related to sustainability in projects by presenting advanced international academic research and practical studies on sustainable project, programme and portfolio management. The results can be adopted into project, programme and portfolio management practices.

The book is structured into three parts. While the first part covers concepts and approaches related to the integration of sustainability in project management, the second part presents research on integrating sustainability into project management in different industries and regions. The final part takes specific perspectives on integrating sustainability into project management related to learning and continuing competence development.

The first part includes the following five chapters.

Chapter 2, by *Lars Nielsen, Philipp Klausing and Peter Nyhuis* is entitled *Towards a Target System to Incorporate Sustainability in Multi-Project Management in Factories.* The authors raise the awareness about the challenge of portfolio balancing, which is requiring an overarching target system for providing guidance. Their research leads to a hierarchy framework according to which sustainability can be integrated and implemented. An exploratory case study, for the German automobile OEMs (Original Equipment Manufacturers) is conducted and the main findings are presented.

In Chap. 3, *Maedeh Molaei, Marcel Hertogh and Marian Bosch-Rekveldt* investigate the *Factors Affecting the Integration of Sustainability in the Early Project Phases in an Integrated Project Management Model.* They conducted a qualitative cross-case analysis on three highways projects in the Netherlands. It is concluded that each project management role is inclined towards specific sustainability dimension. The authors proposed a model for integrating key roles involved in integrating sustainability into project management of infrastructure projects.

Chapter 4 by *Thordur Vikingur Fridgeirsson, Bara Hlin Kristjansdottir and Helgi Thor Ingason* explores the challenges of complexity in the modern business environment and proposes a diagnostic tool, which was designed by the authors to assess the VUCA dimensions a project is facing with. For validation purpose, the tool was tested on different projects from the manufacturing industry. Considering that one of the reasons to integrate sustainability into project management is to reduce project risks, the research is strongly connected with the design of sustainability indicators based on risk management.

In Chap. 5, *Gregory J. Skulmoski, Craig Langston, Alan Patchinga and Amir Ghanbaripour* conceptualize the sustainable approach in the career management. By running an extensive literature review, the authors show how the signals of career risks, such as the absence of substantial training or education, named error messages can be used in a sustainable project-oriented career within the Gig Economy.

Chapter 6 addresses the topic of environmental projects. *Vladimir Obradović, Marija Todorović and Milica Jovanović* discuss the factors that are contributing to or limiting project management application in the field of environment protection at the national and local levels. Based on these factors and the findings of an empirical research conducted in Serbia, the key enablers for successfully integrating project management in environmental projects are identified.

The second part of the book is about integrating sustainability into project management in different industries and regions based on some research results obtained by the authors. This part includes the following ten chapters:

Chapter 7 describes the importance of integrative intelligence for sustainable rural development. By performing an extensive and in-depth analysis of the Mexican rural economy, *Jesus Martinez Almela, Jorge Flores and Jorge J. Peart Mijangos* identify the main types of Rural Mexico and militate for "extensionism", as the founder principle of sustainable rural development. According to the authors, "extension" should cover a wider range of communication and learning activities, organized for rural people and related to different disciplines and subjects, including agriculture, agricultural marketing, health and business studies and project management.

Chapter 8 by Teri Vivienne Okoro is about applying a hybrid framework for reviewing the integration of sustainability in a motorsports tourism project in Nigeria. The applied framework combines the traditional sustainability economic, social, environmental and other dimensions with the co-created benefits and value. This study adopts a qualitative research technique, exploring the challenges of sustainability through interviews relating to a singular case study. Two semi-structured in-depth interviews covered the project context with a more focused examination of project sustainability.

Chapter 9 by Reinhard Wagner is about the process of transforming automotive industry following the adoption of disruptive technologies and sustainable project management approaches. The chapter explains the role of projects in organizing innovation, product development, manufacturing and service delivery and explores the future projectification of this industry. The author made an historical research of the management approaches adopted in the automotive industry, in order to identify the relevant dynamics patterns and future trends.

In Chap. 10, Inga Minelgaite, Bjarnveig Birta Bjarnadottir and Kari Kristinsson present the research on new product development projects in the Icelandic seafood industry regarding three sustainability dimensions, namely environmental, economic and social. Icelandic experience is relevant, considering that Iceland is ranked among the most sustainable countries in the world and the seafood industry in Iceland has been referred to as a role model in the use of fish waste to develop new products. Using survey methodology, the authors investigate how the sustainability dimensions are represented in each phase of new product development projects in the seafood sector.

Chapter 11 is concerned with the *Inter-organizational Co-creation: an Approach to Support Energy Transition Projects*. *Afshin Jalali Sohi, Maryam R. Nezami, Hans Bakker and Marcel Hertogh* argue that restructuring energy systems into more sustainable forms, called "Energy Transition", represents a solution to the actual inefficiency of the energy systems. By conducting a case-study research in The Netherlands, for one energy transition project, the interactions among different actors were mainly investigated in order to better understand to impact of co-creation on the project interactions and on the project success.

In Chap. 12, *Hans Knoepfel* presents a multi-case study on *Sustainability for Construction Projects*. Six large, medium-size and small construction projects were considered in the research. Narrative research was applied in data collection and analysis. In each case, the characteristics about capture, society, economy, environment and resilience were targeted, according to the conceptual model of sustainability of buildings and infrastructure which was proposed by the author, based on the literature review. The initial assumptions were confirmed through the research findings.

In Chap. 13, *Jaehyun Lee, Unho Lee, Eunsang Yoon and Changwoo Park* proposed the usage of Segway model for the assessment of megaproject excellence, especially in the case of Korea's Pilot Smart City Projects. The authors ran several rounds of interviews with project operating groups, mainly decision-makers, how revealed the difficulties in communication and decision-making. It is argued that the usage of the Segway model is a solution for these difficulties. The authors express their interest in continuing the research with the development of the assessment methodology based on the Segway model.

Chapter 14 brings forth the issue of *Public-Private-People Partnership (PPPP) for Infrastructure Development in Indonesia*. The authors *Lukas Beladi Sihombing, Achmad Jaka Santos and Andreas Wibowo* have conducted research for validating the new concept of public-private-people partnership (PPPP). A survey was conducted, mainly for identifying the relevance of people in PPPP. The result showed that the PPPP concept has the ability to benefit and support the livelihood of people living around the project area, thereby ensuring the integration of sustainability into project management.

In Chap. 15, *Sandra Matuhina, Mladen Radujković, Maja-Marija Nahod* present a survey on the role of practice competences in decision-taking process in projects. The survey targeted the challenges of the decision-making, the shift towards the creative leadership and the sustainability of the decision-making. The survey was conducted in three European countries: Croatia, Slovenia and Slovakia.

In Chap. 16, Asa-Bjork Jonsdottir, Haukur Ingi Jonasson, Helgi Thor Ingason and Agnes Holm Gunnarsdottir examine four international business excellence models emphasizing on *social responsibility* and sustainability in order to identify how a company's excellence can be better measured. The applied comparison framework has six dimensions; Leadership, Vision and strategy, Continuous improvement, Processes, People and Universal success and Systematic approach. The authors ran a survey amongst CEOs of 25 Icelandic companies. An important research finding was the actual necessity for the companies to be more oriented than before to the world at large, mainly considering the recent threats, such as global warming and the increase of pollution.

The last section of the book addresses the perspectives of incorporating sustainability in project management through the competence development and includes the following chapters:

Chapter 17 discusses the significance of chatbots in the corporate project management training. *Constanta-Nicoleta Bodea, Maria-Iuliana Dascalu and Alexandru Hang* claim that chatbots are very useful in corporate training, as they support professional development and knowledge sharing to day-to-day business practices, they reinforce learning and help users to find quickly the relevant information, thus contributing to a sustainable education. The authors propose a pattern-based chatbot, useful for corporate training for competence enhancement. The solution was successfully deployed in a professional association and integrated into its digital platform.

In Chap. 18, *Steven Nijhuis* considers that in order to get sustainable teaching of project management, focus on teaching subjects and processes is needed. Comparing the subjects of the offerings of project management, the author concludes that commercial and higher education courses on project management are not only targeting the same group but also give attention to similar subjects. The only significant difference is that higher education tests more the students' knowledge even though there is not a clear proof that learning goals are really achieved. Further research is still required for assuring sustainable project management teaching.

Chapter 19 describes an ex post analysis of the competencies of a team of educators asked to design and implement an online educational programme in a very short track. To achieve this objective, the team applied different methods for crashing the schedule. As a member of the project team, *Adán López Miranda* applied the observation method to collect data for his research. To face these time constraints, the project team members have to develop their project management competences.

Chapter 20 by *Ángela Paneque de la Torre, Vanesa G. Lo Iacono Ferreira, María José Bastante-Ceca and Salvador F. Capuz-Rizo* discusses how certification process organized in Spain during 2014–2019 and based on IPMA ICB standard addressed the competences required to incorporate sustainability in project management of the candidates. Inclusion of sustainability concepts in the project management standard was analysed, together with the assessment tools applied during different stage of the certification programme by Spanish certification body.

Chapter 21 by *Amin Saidoun* addresses the importance of project management competences for the sustainability of non-profit organizations. The empirical investigation was undertaken in two case studies. The first case concerns a medium-sized international fundamental research NPO based in Grenoble (France) and the second case concerns a small national company which advocates the interests of large companies from the logistics sector based in Bonn, Germany.

As editors, we compiled this book with the intention to offer the readers a great variety of approaches and perspectives on sustainability in project management. The book includes relevant regional and sector-specific approaches in integrating sustainability in project management, We are advocating for knowledge co-creation, based on the cooperation between practitioners and academics. For this reason, we consider that this book ensures not only the theoretical grounding but also the practical relevance of the addressed theme.

The book is a valuable resource for researchers and practitioners in project, programme and portfolio management, for educators and trainers as well for post-graduate students, in general, for all those interested to better understand how the sustainability can be integrated into project management. The first part of the book is recommended for scholars interested in general conceptual models and approaches in incorporating sustainability into project management in complex business contexts. Researchers and practitioners will find a great diversity of studies, conducted in different regions and sectors by using a wide range of research methods. The final part of the book is intended to be read mainly by educators and trainers delivering courses and training including as learning topic sustainability in project management.

Thank you for reading!

Part I
Concepts and Models of Integrating Sustainability into Project Management

Chapter 2
Towards a Target System to Incorporate Sustainability in Multi-project Management in Factories

Lars Nielsen, Philipp Klausing, and Peter Nyhuis

Abstract Shortened life cycles and increasing customer-specific mass production confront factories. The professional use of project management is necessary for a permanent adaptation under unstable conditions to retain objective attainment of the factory. Multi-project management is a proven approach to cope with the resulting complex project portfolios. Supported by a literature review, the necessity for a comprehensive and differentiated target system becomes evident, enabling factory management to deliberately position themselves and make decisions in the numerous conflicts of objectives. A predominant contradiction exists between prioritizing projects with short-term profitability while ensuring portfolio sustainability. This challenge of portfolio balancing requires an overarching target system providing guidance. In order to develop this target system, we first developed a hierarchy framework according to which suitable existing approaches are refined and implemented. Furthermore, we illustrate performance indicators to enhance this target system and make it applicable, which we conclusively demonstrate in a case study.

Keywords Multi-project management · Portfolio management · Factory planning · Sustainability · Target systems · Portfolio balancing

2.1 Introduction

Today's factories are characterized by turbulent global markets [20]. The reasons for this are numerous overlapping and mutually influencing factors, which have their origin in particular in the ever-shorter technology cycles, globalization as well as new products and processes [22, 25]. A continuous adjustment of production systems and the yielding increase in the number of projects within factories reflects this dynamic [2, 12, 31]. As a result of the constantly changing environment, factory planning is no longer a one-time task. Rather, factory planning can increasingly

L. Nielsen (✉) · P. Klausing · P. Nyhuis
Institute of Production Systems and Logistics, Leibniz University Hannover,
An der Universität 2, 30823 Garbsen, Germany
e-mail: nielsen@ifa.uni-hannover.de

R. Cuevas et al. (eds.), *Research on Project, Programme and Portfolio Management*,
Lecture Notes in Management and Industrial Engineering,
https://doi.org/10.1007/978-3-030-60139-3_2

be characterized as a permanent process by which production is adapted to the continually changing circumstances [32]. Both proactive and reactive changes are realized through projects and thus affect the total number of existent projects in the factory, which are considered as project landscape. Consequently, a very high project heterogeneity characterizes this project landscape [34].

Existing project management approaches and their consecutive procedures can only insufficiently cope with the increasing complexity within the project landscape of a factory [2, 11]. In addition, little attention is paid to linking higher level company and operational project objectives when selecting projects. As a result, changes in objectives at the company level are not given enough consideration in single projects. Additionally, project sustainability is considered as central criteria for long-term success [1]. Furthermore, a lack of project transparency leads to untapped synergy effects as well as insufficient networking between different projects or possible cumulative risks in the project portfolio. Due to these deficits, systematic and target-oriented multi-project management must supplement conventional single project management in a factory in order to combine effective project selection, managing portfolio sustainability and uncertainties as well as supporting efficient project execution [1, 2, 11, 21, 33].

The hypothesis of this paper, therefore, is that multi-project management has to be enhanced by a target system incorporating all relevant criteria including sustainability, in order to make it applicable in factories. The objectives of factories as well as of multi-project management must be consolidated and coordinated in such a way as to start project initiatives in the factory, which support the collective set of objectives. In order to select, equip and implement those projects, which bring about the change and future sustainable development of the factory in the desired direction, it is necessary to create a common and comprehensive target system. With the help of such a target system, multi-project management decisions in factories can be made in a manner that enables deliberate positioning in prevailing target conflicts. As a consequence, individual projects may be stopped or taken down in priority in terms of resource supply, as long as the overriding attainment of objectives is beneficial.

2.2 Fundamentals

There are always numerous project ideas in companies, various projects are being implemented, some have been cancelled and an even larger number of projects have already been completed and documented. Today, almost all changes in a company are implemented through projects or project-like initiatives [8]. Different project types, objectives pursued, lead times and sizes of the projects characterize the resulting project landscape [15]. Examples for different projects can be long-term reorganizational projects, involving all departments of a company as well as projects to comply with upcoming legal regulations, small improvement projects of everyday workflows or even large factory expansion projects.

Projects initially pursue their individual objectives and should ideally bring about the implementation of company objectives either directly or indirectly. Due to scarce resources and existing dependencies between projects, consequently overall planning and coordination in the form of multi-project management is a reasonable approach [8].

2.2.1 Multi-project Management

Multi-project management primarily comprises the planning and control of the project landscape and provides the necessary organizational and processual framework [8]. Multi-project management therefore is understood as an important function between the strategic level of a company and the operative individual project level [29]. Figure 2.1 shows four relevant elements of multi-project management. The multi-project landscape (a) contains the single projects as well as project programmes, which by itself consist of a number of single projects. The hierarchization (b) of individual projects at the operational level is usually obtained via project programmes and a superordinated project portfolio [7].

The target system (c) is the focus of this paper. It analyses the objectives of the single projects and project programmes and checks their alignment with the higher level factory objectives. Multi-project management deals with the cost-benefit relations of the projects because of the strong influence of the company management and the consideration of their objectives. Thus, it requires positioning in the potential conflict of these objectives [33]. One example of a generic conflict of objectives is the tension between short-term profitability and sustainable projects [1]. Based on the assessment by the target system, a project prioritization (d) can finally be carried out, which allows further actions to be derived, as indicated.

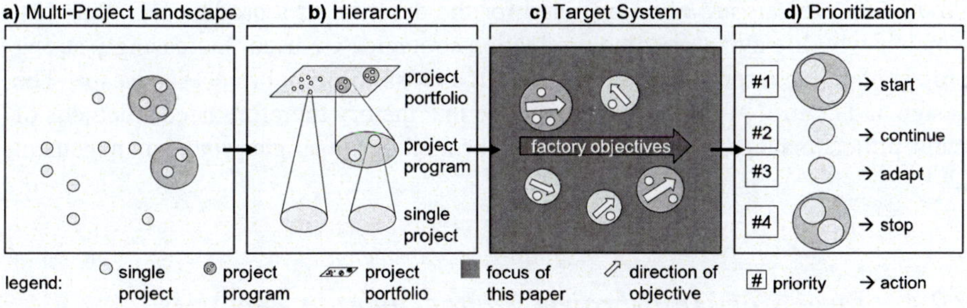

Fig. 2.1 Selection of relevant elements of multi-project management with highlighted focus of this paper, **a** Multi-project landscape, **b** Hierarchy, **c** Target system, **d** Prioritization

2.2.2 Factory Planning

According to the VDI Guideline 5200 [35], a factory is the place of value creation where industrial goods are processed using production factors. The production factors relate to operating resources, materials, personnel, information, capital and space. A factory can therefore be regarded as a socio-technical system in which production systems are formed through the interaction of social and technical subsystems [37].

Factory planning is a subfield of long-term production planning [37]. According to VDI Guideline 5200 [35], it is defined as a target-oriented, structured and systematic process with successive phases, which is carried out using the aid of tools and methods to plan a factory from the definition of factory objectives until the ramp-up of production.

2.2.3 Specificities in Multi-project Landscapes of Factories

Multitudes of different projects characterize the system of a factory. In addition to large and risky factory planning projects (such as expansion projects), equally multilayered IT projects (such as ERP-implementation projects) or smaller organizational projects take place in factories [15]. In addition, projects of the continuous improvement process (CIP) are to be mentioned. Product development projects also play an important role in a factory. Product development is one of the classic application areas of project management, which is continuously applied in research and development. The deliberate transformation of knowledge creates new, further developed or adapted products for the market, which are produced in factories [3].

Due to the strong dynamics in the environment of a factory, the production systems, in particular, have to be adapted regularly. The planning and implementation of such initiatives are usually organized and carried out through projects. Production projects are often operated for the duration of a product life cycle [3], which is why life cycle-oriented production strategies are used. Increasingly strong project planning in production can result in a cross-location project landscape. The design and control of the project landscape in a factory therefore needs methods of multi-project management to supplement the methods of single project management [2, 9].

2.2.4 Targets of Multi-project Management in Factories

By establishing a multi-project management in a company, different objectives of the company can be addressed. According to DIN 69909-1 [8], first of all, transparency is created in the project landscape, thus making connections, synergies and potential risks visible. The single projects and project programmes in the project portfolio should be aligned with the company's targets. The above-mentioned transformations, which have a permanent effect on the factory, can thereby be countered quickly and purposefully. It is therefore necessary to select those projects that should bring about a change in line with the company's objectives. Subsequently, a project prioritization must be carried out, which specifies a ranking order, e.g. for project scheduling and resource management, among the projects in the portfolio that are to be restarted or are already running. Consequently, those project ideas with the highest target conformity can be started or continued with priority and therefore be equipped with the best available resources. The identification and evaluation of opportunities and threats of the individual projects and project programmes also enable portfolio risk management. Furthermore, the projects and project programmes in the project portfolio are regularly monitored for their alignment with the company's objectives and appropriate countermeasures, such as stopping projects or amending their objectives, are initiated in the event of deviations [8].

In factories, the objectives described above also apply but are supplemented by factory specific characteristics. For example, the concrete design of factory objectives is strongly dependent on the type of factory [30]. A low-cost factory, for instance, would place less emphasis on a changeable shop floor with a sustainable air-conditioning concept than an exemplary high-tech factory. In general, factory targets can be divided into formal and factual objectives [10]. Formal objectives are overriding objectives, such as cost-effectiveness and profitability. According to Heger [10], factual objectives contain performance targets, such as quality, logistics performance and changeability.

Linking the various objectives of multi-project management and factory planning in a single model represents a previously unaddressed challenge.

2.3 Factory Requirements for Multi-project Management

The individual conditions of a factory result in different requirements for the most efficient and effective applicability of a target system for multi-project management. This paper only brings forward such requirements, which are particularly important to achieve this purpose.

Planning from rough to detailed results in a hierarchy of objectives that can be used to structure a target system for multi-project management in a meaningful way. Usually, a factory is given overriding corporate objectives, which then have to be broken down into more concrete factory objectives [5, 32]. The project portfolio subsequently is aligned to achieve these factory objectives through approved single project programmes or single projects [11]. A cascading of planning from rough to detailed is therefore necessary since the far-reaching company or factory objectives usually cannot be achieved by singular actions [4, 16, 36].

Promoting portfolio transparency must be aspired by a target system for multi-project management in a factory, in order to systematically plan and control the project landscape. A frequent point of criticism, for example, in factory planning projects, is the lack of interlinking between planning disciplines [27]. Greater transparency allows more interdependencies to be detected and the effects of changes to be represented in a better way [14]. Transparent handling of factory planning projects means that the overall benefit can be maximized, as the individual planning projects can benefit from each other and promote an interaction [24]. Accordingly, an increase in transparency leads to a holistic and targeted cooperation of all projects as well as to an efficient coordination of all members involved [27].

Compliance with project heterogeneity can be derived as requirement originating from the high project variety in a factory. A target system for multi-project management must take this heterogeneity into account and be applicable to all existing types of projects, so that a comprehensive control of the project landscape is possible [8]. Especially the heterogeneity of project clients or sponsors is particularly high [33], whereas the resources involved in the portfolio differ significantly [9].

Compliance with portfolio sustainability in a factory also is an important point that must be considered and represents the focus of this paper. Through their production processes and their considerable influence on the population, factories play a central role in shaping the sustainable development of societies [6]. This requirement therefore addresses a project portfolio in a factory that is ecologically, economically and socially beneficial today and in the future [18]. In order to fulfil this requirement, single project sustainability is important, but has already been discussed extensively [1]. Portfolio sustainability goes beyond taking care of single project sustainability. The sustainability of individual projects is of no gain if the portfolio as a whole is not following the overarching objective of sustainability. A sustainable project portfolio specifically includes the features of 'portfolio balancing' and 'strategic fit', which are described as vital [8, 26]. In the context of the factory, however, there is a particular need for them [2], for example, to start strategically relevant projects that are not profitable in the short term. To consider systematically a certain share of such projects in the portfolio prevents an excessive procrastination in adjustments, which later on would result in enormous correction costs [19].

For a future-oriented factory, *compliance with factory changeability* must also be taken into consideration. A changeable factory can implement structural changes at all levels quickly, with little effort and comparatively low investment expenditures [40]. New products, technologies and markets, as well as fluctuating demand, create a volatile environment to which the factory must adapt quickly and with little effort for a successful existence on the global market [18, 28]. In order to avoid cost-intensive restructuring measures or uneconomical production processes, a target system for multi-project management must meet these requirements and continue to be able to act under rapidly changing conditions.

Compliance with (factory) standards is not to be understood as compliance with legal or other binding requirements, as these are usually covered by existing approaches. Rather, it is intended to comply with voluntary or self-imposed standards or overfulfilment of current regulations for the purpose of expected compliance with future (possibly mandatory) requirements [5].

2.4 Literature Review and Identification of Research Gap

Figure 2.2 compares existing process models with reference to a target system as well as dedicated target systems with the previously mentioned requirements of a target system for multi-project management in a factory.

Although the existing models cover a number of requirements, only a selection of these aspects are in the respective focus of the various approaches. It is also not possible to reach a complete consideration of all requirements by solely combining existing approaches.

Larger deficits can be found in the analysis of existing research with regard to promoting portfolio transparency, compliance with project heterogeneity, as well as compliance with portfolio sustainability. While the first two mentioned are partly covered by some existing approaches, portfolio sustainability is not addressed by any of the assessed models. As a result, no approach currently exists that enables the holistic assessment of the objective attainment of multi-project management decisions in factories with special consideration of portfolio sustainability. Linking the various objectives of multi-project management and factory planning in a coherent model represents a hitherto unsolved challenge. In particular, existing approaches fail to provide a consistent hierarchy of objectives across the different levels to create a coherent target system under inclusion of portfolio sustainability.

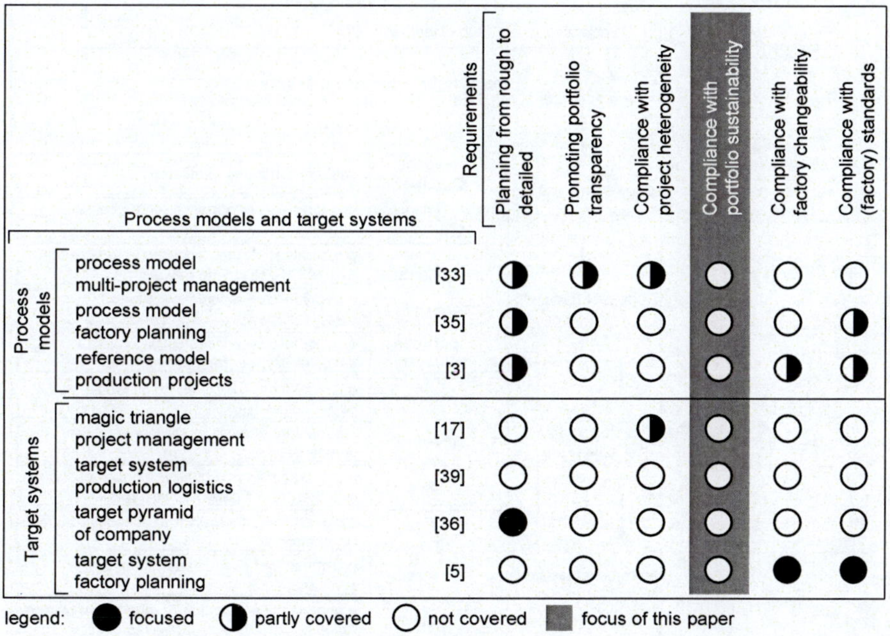

Fig. 2.2 Revealing the research gap by comparing the factory's requirements for multi-project management with existing process models and target systems

2.5 Target System for Sustainable Multi-project Management in Factories

To close the previously mentioned research gap, a corresponding target system, especially for sustainable multi-project management in factories, is presented. It builds upon the findings of the existing approaches already presented. The model is structured according to the hierarchy levels in multi-project management introduced earlier (Fig. 2.1b), supplemented by the levels factory and company [23].

The target system therefore consists of company objectives, factory objectives, project portfolio objectives, project programme objectives and single project objectives. Each of these objectives is to be operationalized by further sub-objectives and KPIs in order to enable a deliberate, target-compliant positioning between the alternatives for action, some of which are conflicting. The classification of objectives, sub-objectives and KPIs in a holistic target hierarchy allows the clear identification of local optimum that could lead to a failure to achieve targets at the higher levels. On the other hand, a local suboptimum can be overcompensated and thus legitimized by a transparent, comprehensible and sustainable global improvement in target achievement. The developed target system for sustainable multi-project management in factories is shown in Fig. 2.3.

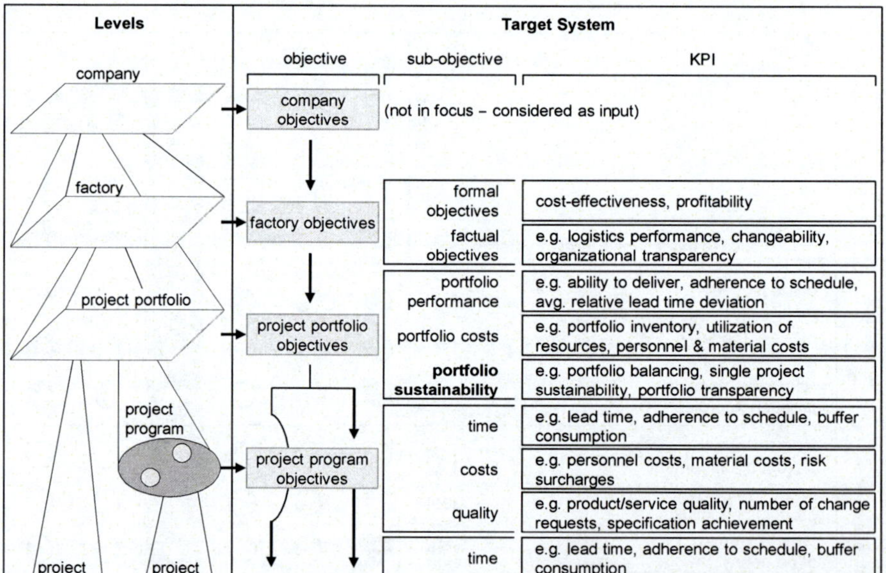

Fig. 2.3 Target system for sustainable multi-project management in Factories

Since all objectives and sub-objectives of the target system can and must be operationalized by various, partly company-specific key indicators, exemplary key indicators are displayed and elaborated in Fig. 2.3 and the following sections. Further information on the applicability of the target system can be found in the corresponding section.

2.5.1 *Factory Objectives*

The factory objectives are influenced by the superordinate company objectives. However, these are outside the focus of this paper and are therefore only considered as a given input. According to Heger [10], factory objectives can be distinguished between formal and factual objectives. The formal objectives include the cost-effectiveness and profitability of a factory. Factory planning must create the framework conditions for an economic and profitable factory operation in order to enable a lasting existence of the factory on the market. The factual objectives are of a rather operational nature and are composed, for example, of logistics performance or changeability. Other factual objectives for example include organizational transparency and organizational interconnectivity [10].

The factory objectives described are used to derive the subordinate project portfolio objectives. In return, the achievement of inferior levels directly influences the achievement of objectives of superior levels.

2.5.2 *Project Portfolio Objectives*

The sub-objectives allocated to the project portfolio objectives consist of portfolio performance, portfolio costs as well as portfolio sustainability. This clustering ensures that the above-mentioned requirements for multi-project management in the factory are taken into account at this level of consideration. It is important to mention explicitly the equal importance and legitimacy of portfolio sustainability as long-term objective alongside performance and cost objectives. Portfolio sustainability therefore is representing the antipole to performance and costs, which together form the magic triangle on portfolio level.

Portfolio performance can be operationalized through several KPIs, including ability to deliver, adherence to schedule and average relative lead time deviation of the project portfolio. These KPIs are used to identify and avoid systematic misplannings or avoidable blockages caused by interdependencies between individual projects that would have a negative impact on project portfolio performance.

The portfolio costs can be determined, for example, by the portfolio inventory or the utilization of resources. Similar to a high inventory level in a warehouse, which results in tied-up capital and other warehousing costs, a high portfolio inventory level of single projects represents a cost driver. These costs are created through a high number of simultaneously running projects with distracting multitasking for all resources. The portfolio inventory should be reduced, which also leads to a positive influence on the average relative lead time deviation (portfolio performance). The utilization of resources should be maximized in order to make the best possible use of available capacities. These KPIs stand in a fundamental conflict of objectives, as utilization of resources and low portfolio inventories are logically contradictory. In addition, other direct costs, such as personnel or material costs should also be considered in portfolio cost observations.

Portfolio sustainability also consists of a number of KPIs. For a comprehensive control of multi-project management, a portfolio balancing between strategic and profitable projects must be considered by compliance with project heterogeneity [13]. A sustainable project portfolio can only be realized persistently through compliance with single project sustainability in the ecological, economic and social areas. In today's turbulent environment, compliance with factory changeability also has a major impact on the sustainability of the project portfolio. Without the project portfolio's ability to adapt quickly to changing conditions, its continued existence is at stake. This changeability is generally described as being facilitated through five changeability enablers [38]. They represent generic approaches to address changeability via modularity, scalability, universality, compatibility and mobility. By emphasizing these five approaches, sustainability can be promoted [40]. The aim should be a balance between efficiency enhancing and changeability supporting projects for the portfolio to be sustainable. Furthermore, a project portfolio requires a maximum of portfolio transparency. Synergies, interdependencies and risks in the project portfolio can only be identified and exploited or overcome through a high degree of openness in single projects, project programmes and the portfolio as a whole.

2.5.3 Project Programme and Single Project Objectives

The objectives of the project programmes, as well as single projects, are derived from the project portfolio objectives. Due to the similar character of project programmes and single projects [8], the same sub-objectives and KPIs can be assumed, which is why this paragraph only refers to the term projects. According to Lock and Wagner [17] a project is subject to three basic target criteria (also called *triple constraint*): time, cost and quality. This *magic triangle* of objectives in project management is an established approach for single projects. The time dimension in a project determines the requirements for set deadlines, measured, for example, through lead time, adherence to schedule and buffer consumption. Another dimension represents costs. All resources used in connection with the project, such as the resulting personnel costs, material costs as well as necessary risk surcharges influence the resulting project costs and need to be considered. The third dimension is the quality of the resulting product, service or subject of a project. This criterion is intentionally very broad and includes all KPIs that are not of a time-based or monetary nature. These include, for example, the product/service quality, the number of change requests or the specification achievements of a project. The objectives contained in the magic triangle are in a permanent conflict so that the improvement of one of the objectives leads to an inevitable deterioration of the other two objectives. This contradiction requires deliberate positioning in the conflict of objectives [17].

2.6 Applicability of the Target System and Case Study

The use of the target system for multi-project management in factories allows a target-oriented alignment of single projects, project programmes as well as project portfolios in order to achieve a high degree of target conformity with the factory and finally the company goals. The presented target system is to be understood as an initial target system to start with following [5]. Based on this universal foundation, it is intended that the factory management supplements or reduces the sub-objectives or KPIs for the individual application case. In addition, a specific weighting of the individual target system components should be carried out during application. In concrete decision-making situations within the project landscape, action alternatives can be analysed and evaluated on the basis of their effects on the KPIs, the sub-objectives and ultimately the objectives.

In our case study, we utilize the situation of German automobile OEMs (Original Equipment Manufacturers) as a unit of analysis. We use an explanatory case (according to [42]) to elucidate the strategic positioning of German automotive OEMs regarding their portfolio balance as a project mix for conventional internal combustion and fully electrical vehicles in their factories. Consequently, we express the overarching research question: How can the strategic decision-making process for project portfolio selection be changed to incorporate sustainability in multi-project management? Furthermore, we formulate the follow-up question: How can the hierarchy framework presented in this paper support the strategic positioning of above-mentioned companies in order to reach higher attainment for sustainability objectives, which are in conflict with performance and cost objectives? The data used for the case study was publically available statements in annual reports.

The current challenge at hand for German automotive OEMs is the uncertain product mixture of conventional internal combustion and fully electrical vehicles in the upcoming years. Drastically diverting scopes of assembly and new regulations, as well as the digitalization of the whole branch, represent various parallel issues for managers [41]. In order to best equip their factories for a hitherto unknown mix of both technologies, the right projects need to be selected for execution. There is a risk that today's automobile manufacturers will prefer the supposedly more profitable short-term improvement projects for increasing the output of conventional combustion vehicles. This would result in shortcomings of the strategically relevant, but very capital-intensive and partly less efficient assembly systems, which offer a high degree of changeability with regard to the combustion/electrical mix to be expected in the future. Conventional portfolio management would select projects according to the individual project objectives (time, cost, quality). Possibly they also consider performance and cost objectives on a portfolio level, but certainly do not take sufficient account of project portfolio sustainability. Due to the equal positioning of portfolio sustainability alongside the performance and cost targets as suggested in this paper, a deliberate decision against a sustainable project at least requires elaborated justification or legitimation. This enables appropriate portfolio balancing in accordance with all relevant criteria.

The main findings of the case study consequently are the following. German automotive OEMs would normally favour projects in their prioritization, which solely support vehicles with internal combustion engines. However, through the utilization of the target system presented in this paper, the project selection process would increase the importance of sustainability by placing it on the same level as performance and cost indicators. The research question therefore can be answered with this explanatory case. Nevertheless, further research with regard to a broad empirical study is recommended to substantiate the results determined in this case.

2.7 Conclusion

In today's factories, the professional use of multi-project management is unavoidable due to rapidly changing conditions. It has been shown that planning and controlling the project landscape in a factory requires a comprehensive and factory-specific target system. The special requirements of a factory for such a target system for sustainable multi-project management in factories were presented in detail. A literature review showed that the existing approaches from the fields of process models or target systems only insufficiently take into account the identified requirements. Thus, a research gap could be identified in the form of a target system for sustainable multi-project management in factories.

This paper therefore contributes to the creation of such a target system addressing the specific requirements of a sustainable multi-project landscape in factories. In order to derive the objectives of the project portfolio and consequently the project programmes and single projects from the factory and company objectives, the target system is first hierarchically structured into five levels. Subsequently, sub-objectives were assigned, which allow a breakdown and thus a more differentiated view. Finally, exemplary KPIs were assigned to each sub-objective that build upon the existing approaches and take into account the identified special requirements of the factory as well as making their achievement measurable at each level. Finally, the necessary steps for the applicability of the target system were explained.

To conclude, it can be stated that with the target system for sustainable multi-project management in factories an essential building block for the consideration of the special requirements of a factory was developed. Future research work and publications of the Institute of Production Systems and Logistics will work towards the development of a comprehensive and integrated process model for multi-project management in the factory. This will enable managers in the factory to carry out the numerous tasks and associated decisions within the multi-project management of a factory effectively and efficiently.

References

1. Aarseth W, Ahola T, Aaltonen K et al (2017) Project sustainability strategies: a systematic literature review. Int J Project Manag 35(6):1071–1083. https://doi.org/10.1016/j.ijproman.2016.11.006
2. Aurich JC, Barbian P, Naab C (2005) Multiprojektmanagement in der projektorientierten Produktion. Gestaltung und Lenkung der Projektlandschaft in der Produktion. wt Werkstattstechnik online 95(1/2):19–24
3. Barbian P (2005) Produktionsstrategie im Produktlebenszyklus. Konzept zur systematischen Umsetzung durch Produktionsprojekte. Zugl.: Kaiserslautern, Techn. Univ., Diss., 2005. Produktionstechnische Berichte aus dem FBK, 01/2005. Techn. Univ, Kaiserslautern
4. Bea FX, Haas J (2017) Teil 2: Strategische Planung. In: Bea FX, Haas J (eds) Strategisches management. UVK Verlagsgesellschaft mbH, Konstanz, pp 46–241
5. Brieke M (2009) Erweiterte Wirtschaftlichkeitsrechnung in der Fabrikplanung. Berichte aus dem IFA, vol 2009,1. PZH, Produktionstechnisches Zentrum, Garbsen
6. Bundesministerium für Umwelt, Naturschutz und Reaktorsicherheit Umweltbericht 2010. Umweltpolitik ist Zukunftspolitik (2010) https://www.bmu.de/fileadmin/Daten_BMU/Download_PDF/Strategien_Bilanzen_Gesetze/umweltbericht_2010_bf.pdf. Accessed 30 Jul 2018
7. DIN Projektmanagement: Projektmanagementsysteme - Teil 5: Begriffe (69901-5) (2009)
8. DIN Multiprojektmanagement: Management von Projektportfolios, Programmen und Projekten—Teil 1: Grundlagen (69909-1) (2013)
9. Dombrowski U, Ernst S, Boog H (2015) Multiprojektmanagement in der Fabrikplanung. Integration des Multiprojektmanagements in die Umplanung von Fabriken unter Einsatz von virtuellen Teams. Industrie 4.0 Management 31(4):43–47
10. Heger CL (2007) Bewertung der Wandlungsfähigkeit von Fabrikobjekten. Berichte aus dem IFA, vol 2007, 1. PZH, Produktionstechnisches Zentrum, Garbsen
11. Hiller M, Klusch M, Monjé M (2001) Multiprojektmanagement als Führungsinstrument der Zukunft. ZWF 96(6):317–321. https://doi.org/10.3139/104.100434
12. Hiller M, Barbian P, Warnecke G et al (2002) Projektorientierung der Produktion. ZWF 97(10):514–518. https://doi.org/10.3139/104.100577
13. Hope AJ, Moehler R (2014) Balancing projects with society and the environment: a project, programme and portfolio approach. Procedia—Soc Behav Sci 119:358–367. https://doi.org/10.1016/j.sbspro.2014.03.041
14. Kampker A, Burggräf P, Krunke M et al (2014) Das Aachener Fabrikplanungsvorgehen. Agile Fabrikplanung im turbulenten Umfeld. wt Werkstattstechnik online 104(4):192–196
15. Kunz C (2007) Strategisches Multiprojektmanagement. Konzeption, Methoden und Strukturen, 2nd edn. Unternehmensführung & Controlling. Deutscher Universitäts-Verlag GWV Fachverlage GmbH, Wiesbaden (2007)
16. Kuster J, Bachmann C, Huber E et al (2019) Handbuch Projektmanagement. Springer, Berlin, Heidelberg
17. Lock D, Wagner R (eds) (2019) The handbook of project portfolio management. A Gower book. Routledge Taylor & Francis Group, London, New York
18. Loos MN, Ovtcharova J, Heinz S (2012) Fabrikplanung im Fokus. Im Spannungsfeld zwischen Anlagen und baulichen Strukturen. Bautechnik 89(4):257–263

19. Lübkemann, J.: Ermittlung des Restrukturierungsbedarfs von Fabriken. Dissertation. Berichte aus dem IFA, 2016, Band 9. PZH, Produktionstechnisches Zentrum, Garbsen
20. Mack O, Khare A (2016) Perspectives on a VUCA World. In: Mack O, Khare A, Krämer A et al (eds) Managing in a VUCA World. Springer International Publishing, Cham, pp 3–19
21. Martinsuo M, Korhonen T, Laine T (2014) Identifying, framing and managing uncertainties in project portfolios. Int J Project Manag 32(5):732–746. https://doi.org/10.1016/j.ijproman.2014.01.014
22. Nielsen L, Schmidt C, Blume S et al (2016) Towards quantitative factory life cycle evaluation. Procedia CIRP 55:266–271. https://doi.org/10.1016/j.procir.2016.08.009
23. Nielsen L, Klausing P, Nyhuis P (2019) Zielsystem für Multiprojektmanagement in Fabriken. Wie zahlreiche, parallele Fabrikprojekte an übergeordneten Zielvorgaben ausgerichtet werden können. wt Werkstattstechnik online 109(4):273–277
24. Nyhuis P, Elscher A, Kolakowski M (2004) Prozessmodell der Synergetischen Fabrikplanung. Ganzheitliche Integration von Prozess- und Raumsicht. wt Werkstattstechnik online 94(4):95–99
25. Nyhuis P, Heinen T, Reinhart G et al (2008) Wandlungsfähige Produktionssysteme. Theoretischer Hintergrund zur Wandlungsfähigkeit von Produktionssystemen. wt Werkstattstechnik online 98(1/2):85–91
26. Rabl W (2013) Projektmanagement. In: Kleinaltenkamp M, Plinke W, Geiger I (eds) Auftrags- und Projektmanagement. Mastering Business Markets, 2., vollständig überarbeitete Auflage 2013. Springer Gabler, Wiesbaden, pp 301–381
27. Reinema C, Pompe A, Nyhuis P (2013) Agiles Projektmanagement. Einsatzpotenziale und Handlungsbedarfe im Rahmen von Fabrikplanungsprojekten. ZWF 108(3):113–117. https://doi.org/10.3139/104.110903
28. Reinema C, Schmiedgen P, Noennig J et al (2013) Kommunikationsstrukturen in Fabriken. Konzept zur Gestaltung dynamischer Geschäfts- und Kommunikationsprozesse in Fabriken. wt Werkstattstechnik online 103(4):269–273
29. Reinhart G, Zäh MF (2003) Marktchance individualisierung. Springer, Berlin Heidelberg, Berlin, Heidelberg
30. Schenk M, Wirth S, Müller E (2014) Fabrikplanung und Fabrikbetrieb. Methoden für die wandlungsfähige, vernetzte und ressourceneffiziente Fabrik, 2., vollst. überarb. und erw. Aufl. VDI-Buch. Springer Vieweg, Berlin, Heidelberg
31. Schoper Y-G, Wald A, Ingason HT et al (2018) Projectification in Western economies: A comparative study of Germany, Norway and Iceland. Int J Project Manag 36(1):71–82. https://doi.org/10.1016/j.ijproman.2017.07.008
32. Schuh G (2007) Fabrikplanung im Wandel. Gibt es die ideale Fabrik? wt Werkstattstechnik online 97(4):194
33. Seidl J (2011) Multiprojektmanagement. Übergreifende Steuerung von Mehrprojektsituationen durch Projektportfolio- und Programmmanagement. Xpert.press. Springer, Berlin, Heidelberg
34. Terkaj W, Tolio T (2019) The Italian flagship project: factories of the future. In: Tolio T, Copani G, Terkaj W (eds) Factories of the future, vol 20. Springer International Publishing, Cham, pp 3–35
35. VDI: Fabrikplanung Planungsvorgehen (5200 Blatt 1) (2011)
36. Weber W, Kabst R, Baum M (2014) Einführung in die Betriebswirtschaftslehre, 9., aktual. und überarb. Aufl. Gabler, Wiesbaden
37. Westkämper E (2013) Digitale Produktion—Einführung. In: Westkämper E, Spath D, Constantinescu C et al (eds) Digitale Produktion. Springer, Berlin, Heidelberg, pp 1–16
38. Wiendahl H-P, ElMaraghy HA, Nyhuis P et al (2007) Changeable manufacturing—classification, design and operation. CIRP Ann Manuf Technol 56(2):783–809. https://doi.org/10.1016/j.cirp.2007.10.003
39. Wiendahl H-P, Wiendahl H-H (2019) Betriebsorganisation für Ingenieure, 9., vollständig überarbeitete Auflage

40. Wiendahl H-P, Reichardt J, Nyhuis P (2015) Handbook factory planning and design. Springer, Heidelberg
41. Winkelhake U (2018) The digital transformation of the automotive industry. Springer International Publishing, Cham
42. Yin RK (2018) Case study research and applications. Design and methods, Sixth edition. Sage, Los Angeles, London, New Delhi, Singapore, Washington, DC, Melbourne

Chapter 3
Factors Affecting the Integration of Sustainability in the Early Project Phases in an Integrated Project Management Model

Maedeh Molaei, Marcel J. C. M. Hertogh, Marian G. C. Bosch-Rekveldt, and Robin Tamak

Abstract This study investigates the factors affecting the integration of sustainability into the project management of infrastructure projects, specifically highway projects during early phases. The research was drawn upon previous studies in order to develop a sustainability framework for measuring the project success in three aspects of sustainability: People, Planet, and Prosperity (triple bottom line). Next, Critical Success Factor (CSF) framework in the construction sector was extracted through a comprehensive literature review. A qualitative cross-case analysis was conducted on three sustainability-oriented highways projects in the Netherlands. Data were collected through document review and twelve in-depth interviews with different roles of Integrated Project Management (IPM) model. The findings suggest that each IPM role is inclined towards specific sustainability dimension which affects the application of sustainability CSFs. The results reveal that among the sixteen identified CSFs promoting the integration of sustainability, following factors were acknowledged by all the IPM roles: awareness of project external factors, clearly defined scope, clearly defined goals/ambitions. Further, the paper conceptualizes a model for integrating key roles involved in the project management of infrastructure projects. The model is based on the triple bottom line of sustainability bringing all the roles involved in the project management of infrastructure projects together.

M. Molaei (✉) · M. J. C. M. Hertogh · M. G. C. Bosch-Rekveldt · R. Tamak
Faculty of Civil Engineering and Geosciences, Delft University of Technology, Stevinweg 1, Postbox 5048, 2600 GA Delft, The Netherlands
e-mail: Maedeh.molaei@tudelft.nl

M. J. C. M. Hertogh
e-mail: M.J.C.M.Hertogh@tudelft.nl

M. G. C. Bosch-Rekveldt
e-mail: M.G.C.Bosch-Rekveldt@tudelft.nl

R. Tamak
e-mail: Robin.Tamak@gmail.com

R. Cuevas et al. (eds.), *Research on Project, Programme and Portfolio Management*, Lecture Notes in Management and Industrial Engineering,
https://doi.org/10.1007/978-3-030-60139-3_3

Keywords Project management · Sustainability · Infrastructure projects · Highway projects · IPM model · CSFs

3.1 Introduction

Recently, the concept of sustainability has evolved as one of the challenges and focal points for society [1]. Sustainability can be regarded as an important project goal which entails a broad range of value and benefits [2]. The focus on sustainability has resulted into the emergence of studies on the integration of this concept into project management practices suggesting that sustainability can be considered as new school of thought in project management [3, 4]. This school of thought mainly includes the following features: considering projects in a societal perspective, having a 'Management for stakeholders' approach, applying Triple bottom line criteria, and taking a value based approach to projects and project management [4].

Specifically in the construction sector, companies are criticized for their conventional approach focusing only on their short-term benefit [5]. In addition, construction industry is considered as one of the main polluters of the natural resources such as carbon emissions, air and water quality [6]. Thus, there is more pressure on these companies to extend their accountability and focus more on social and environmental aspects of their business [7]. The transport sector is assumed to be the second largest emitter of carbon dioxide (CO_2) emissions, the main anthropogenic greenhouse gas [8]. The emissions from the transport sector might double by 2050 due to the fast development of the emerging economies [9]. Highway projects, thus, play a significant role in bringing a change and create value into society by integrating sustainability. This is also very important since by 2030, the public project organization of Dutch highway projects aims to become energy neutral and working according to the circular principle [10].

The current study contributes to both science and practice. Regarding the scientific contribution, the study suggests a framework of the sustainability success factors. Sabini et al. [11] performed an extensive and systematic literature review of 770 publications focused on sustainability and project management from the period 1993 to 2017. They identified three themes showing different views on sustainability: (1) The value of implementation of sustainability into project management (why); (2) The extent to which sustainability affects traditional project management practices (what); and (3) The determinants for the optimal implementation of sustainable project management (SPM) and providing practical suggestions (How). Following this explanation, the current study positions itself in the third identified theme of sustainability literature by providing recommendations on how sustainability can be implemented in highway projects. Regarding the practical implications, the developed model in this research can be used by managers in infrastructure projects oriented towards sustainability. This paper is based on Tamak [12]'s Master thesis on integrating sustainability into project management.

The main objective of this research is to inspect sustainability in the current project management practice and propose a conceptual model which helps improving project success of a sustainability-oriented highway project. More specifically, this research is aimed to prepare current IPM managers of a public project delivery organization for future highway projects to become more sustainable-oriented. Based on the aforementioned problem statement and the objective of this research, the research question is formulated as:

> How can Critical Success Factors be applied in an integrated project management model to improve the chances of project success during the exploration and planning phase of a sustainability-oriented highway project?

The paper is structured as follows. The next section acknowledges sustainability in project management literature as a project success criterion and relevant Critical Success Factors in the construction industry. Then, the methodology for empirical data gathering is explained followed by the research findings. After the discussion, the conclusions and potential directions for further research are given.

3.2 Literature Review

3.2.1 Sustainability as a Project Success Criterion

Project success is a multi-dimensional concept [13, 14]. This includes the project efficiency, impact on the project team, impact on the customer/client, business and direct success, and preparation for the future. However, more recent literature also incorporates environmental, economic, and social sustainability dimensions [15]. Currently, it is widely accepted that achieving sustainability ambitions becomes crucial for measuring the overall success of infrastructure projects [16]. Carvalho et al. [15] conducted a survey in a wide range of industrial sectors, countries, project complexity, and project size. Their findings show a significant relation between project success and project social & environmental impacts with project sustainability management. Thus, findings from literature suggest use of sustainability as an upcoming project success criterion.

3.2.2 Definition of a Sustainable Highway

In order to explain how sustainability can be defined in a highway project, Elkington's People Planet Profit (3P) principle was selected as a sound sustainability theory [17]. People and Planet dimensions of sustainability have remained unchanged in the recent literature. With regard to third pillar, however, recent literature suggests "prosperity" as a concept that goes beyond economic development, i.e. Zimmerman

[18]. The Council of Transport Ministers of the European Union adopted a definition for a sustainable transport as a system Windhoff-Héritier et al. [19]. Hence, based on these explanations, in this study, the following definition for sustainable highway is considered:

- For People: allows the basic access and development needs of individuals, stakeholders and societies involved in the surroundings to be met equitably and in a manner consistent with the ecosystem
- For Planet: ensures environmental protection or limits the impact on the ecosystem while consideration of economic feasibility and the society (stakeholders) involved
- For Prosperity: is affordable, operates fairly and offers added value to support a competitive and balanced economy in the long term

In order to establish sustainability as a project success criterion, a distinct set of success sub-criteria is required. For developing a distinct sustainability success sub-criteria list, recent literature is reviewed. Gijzel et al. [20] developed a framework for sustainable aspects of a tunnel. We adopted this framework to identify sustainability success sub-criteria of a highway. Based on three expert judgments and discussions [12], a total of 30 sustainability success sub-criteria for a highway project, were validated and defined for this study (see Appendix 1).

3.2.3 Success Factors for Implementing Sustainability

Some earlier studies investigated Critical Success Factors (CSFs) for integrating sustainability into project management practices (i.e. Mavi et al. [21]; Martens et al. [22]. Another attempt was made by Banihashemi et al. [23] to review the recent literature and identify CSFs for properly integrating sustainability into project management practices of construction projects in developing countries. Specifically, following Slaughter [24], they identified success factors at five different stages of identification, evaluation, commitments, preparation (on projects and in organization) and implementation. Despite these earlier studies, there is still call for practical implications into how sustainability has been emended into project management.

The current study builds upon earlier work of Molaei et al. [25] on identifying factors leading to the project success through an extensive literature review. The modified list of factors clustering into seven categories, following Westerveld [26]. Recognition of the success causes or failure cause is crucial for maintaining the sustainability of infrastructure projects, which are usually publicly funded. A project delivery organization is a separate entity within a public domain, responsible for delivering the project and in essence functioning as client towards contractors [27]. The role of the client for integrating sustainability is crucial, since one of the influential enablers for considering sustainability is whether "the client asks for it" [28]. The current research is performed in a public project delivery organization in the Netherlands which is an agency of the Ministry for Infrastructure and Environment.

This organization is responsible for project success on behalf of the public client. Currently, this organization follows an Integral Project Management (IPM) model for performing projects, consisting of five managers or roles, who can dominantly affect project success of any undertaken project which is further elaborated in Sect. 3.3.2. It is evident that IPM roles carry out CSFs through their professional role, which shows the demand for this research.

3.3 Research Method

3.3.1 Research Design and Case Selection

As this study aims to explore the recent nature of sustainability as a project success criterion and to identify CSFs by in-depth investigation of sustainability-oriented highway projects, a case study approach was followed. Multiple case studies help in providing deeper understanding to the researcher. Knowledge obtained from different cases can be used to generalize across similar cases [29]. The aim of this study is to see processes and outcomes across the cases, to understand how these processes are qualified by local conditions, and thus to develop more sophisticated descriptions and powerful explanations. Case studies provide means to verify findings from literature [30].

Two criteria were set for selecting the cases: (1) The highway project should be sustainable-oriented in the public project delivery organization, (2) The exploration and planning phase of the project should be completed or near completion. This enables access to the IPM roles or other practitioners working on the project.

3.3.2 Interview Data and Analysis

The primary sources of data for a case study "*comes mostly from document reviews, interviews, observation, and secondary analysis*" [31]. Extensive document review was conducted to identify sustainability themes (or) goals (or) ambitions of the three selected highway projects and subsequently look for used CSFs. The document review of the highway projects was done prior to conducting interviews. This was done to verify or clarify the findings from the case studies.

Next, the semi-structured interviews were conducted. The purpose of the research and the anonymity were explained to the interviewees before conducting the interviews. Based on the availability of IPM roles in each case, four respondents per case have been selected for the interviews. The roles of the respondents per case are presented in Appendix 2. As discussed, the IPM model consists of five and sometimes more roles, depending on the project size and complexity. The five fundamental roles include Project Manager (PM), Project Control Manager (PCM), Project Environment Manager (PEM), Technical Manager (TM), and Contract Manager (CM). For each case, next to the project manager, at least two different respondents, who represent one fundamental IPM role, are involved. The two extra IPM roles are assumed to provide enough experience and expertise for comparison with other fundamental IPM roles.

Each interview consists of three parts. (Part 1): First the respondents were asked to judge to what extent the validated 30 sustainability success sub-criteria were perceived as important in their specific highway project. They explicitly asked to rank these sub-criteria on a scale from 1 to 5. (Part 2): The second part of the interviews focused on the semi-structured interviews where open-ended questions were asked for the identification of CSFs for the implementation of sustainability. (Part 3): Finally, the respondents were requested to verify the list of 28 CSFs by ranking them from 1 to 5 and to express if any CSF was missed during the interview.

All interviews were recorded and transcribed while maintaining the anonymity of the interviewees and the cases. The identified CSFs were coded by a combination of inductive and deductive approaches. In deductive coding, existing theoretical framework of CSFs was used for identification of CSFs in the case studies. Inductive coding entails the identification of new CSFs to be acknowledged by the respondents.

3.4 Results

The results of the three parts of the interviews were analyzed in three stages. In stage 1, CSFs which is presented in all the cases were identified by following this approach: the frequency of occurrence of each of the identified CSFs were determined. If a specific CSF was stated by the majority of the respondents (75% or more), it was considered for further data analysis. Then, the codes are combined to develop the final Sustainability Success Factors (SSFs) framework as presented in Table 3.1.

In stage 2, the inclination of the IPM roles to sustainability dimensions (People, Planet, Prosperity) was determined by analyzing the results of the ranking of sustainability sub-criteria during part 1 of the interviews. The findings suggest that not all IPM roles have equal inclination towards sustainability dimensions.

The PM is responsible for various aspects in the project having a balanced view of sustainability dimensions. The PCM role is not affected by sustainability making this role to work outside the dimensions of sustainability. The PEM, TM and CM are directly connected and constantly involved with various stakeholders, design/technical teams and contractors respectively. Based on their ranking of the sustainability sub-criteria, the PEM is positioned mainly in the People dimension, the CM is positioned in the Prosperity dimension, and the TM can be placed in the Planet dimension.

The last stage includes determination of the inter-relationships between the inclination of the IPM roles towards sustainability dimensions, and use of common SSFs across the cases. These patterns are based on specific type of interaction between the

Table 3.1 Sustainability Success Factors (SSFs)

No.	SSFs	Definition
1	Awareness of project external factors	The awareness of project regarding sustainability goals/ambitions with respect to policy, society, technology and economic context
2	Clearly defined scope	A clear, well defined scope for sustainability goals/ambitions through establishment of boundaries and constraints (standards) & acknowledgement of ambitions by the client
3	Information sharing within the project team	Use of timely (active) distribution of necessary and valuable information regarding sustainability goals/ambitions through efficient communication channels from different project parties within the project team
4	Monitoring & Control	Use of standard control and monitor mechanisms for sustainability goals/ambitions through detailed plan, change management process, inspection/supervision and feedback mechanism to ensure acceptable progress on time, cost and scope
5	Risk Management	Use of risk oriented warning system and risk sessions to identify, define, analyse and assess risks pertaining to sustainability goals/ambitions
6	Proper selection of contracting strategy/tender process	Use of an adequate contracting strategy and competitive tender process that incorporates and promotes sustainability goals/ambitions (explicitly states sharing of risks and clarity about responsibility)
7	Collaboration between project parties	Required level of collaboration/cooperation among project participants for definition and implementation of sustainability goals/ambitions through an open (positive) attitude and effective communication
8	Competent/multidisciplinary team	Use of a suitably qualified project team to define and achieve sustainability goals/ambitions
9	Top management support	Commitment of senior management of the organization for the sustainability goals/ambitions
10	Client involvement	Timely consultation of client for decisions and support regarding sustainability goals/ambitions

(continued)

Table 3.1 (continued)

No.	SSFs	Definition
11	Active involvement of stakeholders	Timely involvement of stakeholders, in various stages of project to improve commitment, provide continuous support, lay grounds for negotiations, minimize opposition, develop mutual trust and improve overall communication for the sustainability goals/ambitions
12	Clear goals & Ambitions	Clear goals (obligatory) & ambitions (that have added value) regarding sustainability, linking to the requirements of the client
13	Sustainability policy	Use of sustainability policy
14	Systematic planning	Use of a realistic and detailed project plan to achieve the sustainability goals/ambitions
15	Adequacy & Efficiency of resources	Presence of available and competent resources for achieving sustainability goals/ambitions
16	Affinity for sustainability	Presence of ambitious nature, personal drivers and like-mindedness within the project team for sustainability goals/ambitions

two variables, namely inclination of the IPM roles and the SSFs [30]. The results of this analysis are plotted in a matrix as presented in Table 3.2. In this table, each “X” represents evidence from the interviews in at least two cases, to justify the relationship between two variables.

The results suggest that not all the IPM roles support application of all SSFs in order to improve chances of project success of a sustainability-oriented highway project. Thus, it can be observed that there might be specific focus area for each single IPM role. For instance, application of proper risk management for achieving sustainability ambitions is merely the responsibility of PCM. Active involvement of stakeholder is also acknowledged to be the responsibility of the PEM. In addition, the findings indicate the interdependency between the IPM roles to achieve a successful sustainable project. PM is considered to select the competent project team for implementing sustainability. As another example, defining a clear project scope for sustainability ambitions and determining the boundaries for is jointly attributed to all the IPM roles. Finally, CM is responsible for selecting a contracting strategy where sustainability is considered as selection criteria.

3.5 Discussion

The results from the cross-case analysis suggest that there is different inclination of the IPM roles towards sustainability dimensions. The results of the literature review and the empirical investigation are synthetized to propose a conceptual model for

Table 3.2 Matrix of the interrelationships between the inclination of IPM roles towards sustainability dimensions, and use of SSFs

	SSFs	PEM (people dimension)	TM (planet dimension)	CM (prosperity dimension)	PM (3P)	PCM (no dimension)
1	Awareness of project external factors	×	×	×	×	×
2	Clearly defined scope	×	×	×	×	×
3	Information sharing within the project team					×
4	Monitoring & Control					×
5	Risk management					×
6	Proper selection of contracting strategy/tender process			×		
7	Collaboration between project parties	×	×	×		
8	Competent/multidisciplinary team				×	
9	Top management support				×	
10	Client involvement				×	
11	Active involvement of stakeholders	×			×	
12	Clear goals & Ambitions	×	×	×	×	×
13	Sustainability policy	×	×	×	×	
14	Systematic planning					×
15	Adequacy and efficiency of resources				×	×
16	Affinity for sustainability	×	×	×	×	

integrating sustainability into project management practices. The model is called Integrated People, Planet, and Prosperity Management or I3PM model consisting of five fundamental roles, namely, Project Manager, Project Control Manager, Project Environment Manager, Technical Manager, and Contract Manager (see Fig. 3.1). This conceptual model depicts the focal points of the IPM roles and integrates the five roles to deliver a sustainable infrastructure project. Three of these roles, namely PEM, TM and CM, work in sustainability dimensions of People, Planet and Prosperity, respectively. PM is positioned in the intersection of the three dimensions, reflecting this crucial role in implementing sustainable ambitions of a highway project. The fifth role, PCM, is positioned outside the sustainability dimensions acting as the coordinator who has inter-dependency on other IPM roles to carry out his professional role and vice-versa.

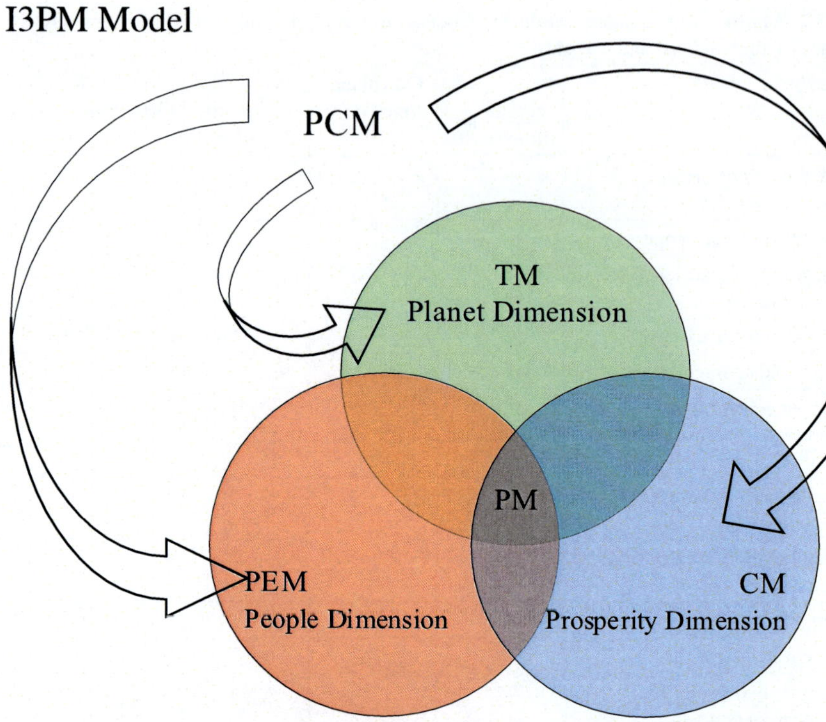

Fig. 3.1 Integrated people planet prosperity management model. *Note* PM stands for Project Manager; PCM for Project Control Manager; PEM for Project Environment Manager; TM for Technical Manager; and CM for Contract Manager

The model also consists of sixteen SSFs, which were identified from the relationship between the inclination of IPM roles towards sustainability dimensions and use of SSFs as explained in Sect. 3.4. The model and the list of SSFs were further validated through an expert judgement within the public project delivery organization. Experts confirmed that the application of SSFs could improve the chances of project success. However, based on the experts' views, two adjustments were made with respect to the boundaries of I3PM model. First, clearly defined scope is modified based on the expert comments to "Flexible scope" which is defined as "*a flexible scope for sustainability goals/ambitions through establishment of opportunity space by the client*". This finding also supports the study of Lechler et al. [32] suggesting that identifying opportunities for maximizing the project value requires a flexible approach which might result in changing the scope.

Second, affinity towards sustainability was acknowledged by the respondents as a new success factor for implementing sustainability in infrastructure projects. This is also confirmed by the study of Silvius et al. [33] where they claim that the extent to which sustainability is considered, depends on the project manager's personal attitude and training. Based on the findings, they distinguished three groups of project managers [33]:

1. Pragmatic managers, who integrate sustainability upon good applicability and they are not self-motivated for sustainability;
2. Intrinsically motivated managers, who care about the nature and their environment and this behavior is self-stimulated;
3. Task-driven managers, who consider sustainability only if it is part of the project's requirements or objectives.

This SSF is allocated to IPM roles which are professionally affected by sustainability, namely PM, PEM, TM and CM. However, the corresponding position of this SSF (with regard to IPM roles) is changed. Based on the perspectives of the experts, sustainability cannot be forced and it should be indistinctly embedded in an IPM role. The SSF "competent/multidisciplinary team" was mainly agreed as a SSF under PM role whereas "affinity for sustainability" is strongly suggested to be part of other IPM roles as well. The reason might stem from the fact that a PM can initiate the implementation of sustainability by appointing a competent/multidisciplinary team which is in line with the findings of Silvius et al. [33]. However, maintaining the affinity of sustainability requires collective approach of all IPM roles since it depends on personal nature of all the team members and NOT on leadership/team building skills of a PM.

The final list of SSFs and the corresponding IPM role responsible for implementing them to be used in the I3PM model is summarized in Table 3.3. This provides a practical approach for implementing and integrating sustainability into project management practices.

3.6 Conclusion

The paper adopted a qualitative approach in which the findings of the cross-case analysis of three projects revealed sixteen Sustainability Success Factors for achieving the sustainability ambitions of highway projects. Application of these SSFs are deemed to improve chances of identified sustainability success sub-criteria of a sustainability-oriented highway project. Next, SSFs are synthesized and integrated into an I3PM model identifying various SSFs to be implemented by each of the roles in this model. I3PM model could help project management practice for better achieving sustainability goals/ambitions of a highway project. The conceptual model is further validated through expert judgement. This model provides a direction to focus on crucial areas during exploration and planning phase of a sustainability-oriented highway project.

The use of qualitative cross-case analysis can be considered as one of the limitations of this study. However, by doing the expert judgement, the credibility of the results, and thus, the possibility of generalizability of the findings is increased. Another limitation was the use of the triple bottom-line sustainability dimensions as underlying sustainability theory which forms the sustainability sub-criteria. Other aspects of sustainability such as lifecycle thinking was not explicitly considered in this project which can provide some future directions for the research.

All projects in this study were performed through DBFM (Design, Build, Finance, and Maintain) contract where the project lifecycle is considered by making the contractor responsible for a longer duration due to "Finance and Maintenance" aspects. Use of DBFM or DBFM plus, and the use of Innovative partnerships are suggested as ways to include sustainability success sub-criteria in a contract, supporting the study of Kivilä et al. [2] suggesting that alliance contract activates the partners to exploit innovation opportunities.

In addition, sustainability is highly context dependent [34] which suggests further research into inclusion of the contextual factors in the model. This research proposed the I3PM model within a public project delivery organization (predominantly based on a client's perspective). Contractor's perspectives need to be investigated as well.

Compliance with Ethical Standards

Table 3.3 SSFs and the corresponding IPM role responsible for implementing them

IPM role responsible for implementing SSFs	SSFs
All IPM roles	Awareness of project external factors
	Flexible scope
	Clear goals/ambitions
PM/PEM/TM/CM	Sustainability policy Affinity for sustainability
PEM/TM/CM	Collaboration between project parties
PM/PCM	Adequacy & Efficiency of resources
PM	Competent/multidisciplinary team
	Top management support
	Client involvement
PCM	Information sharing within the project team
	Monitoring and Control
	Risk Management
	Systematic planning
CM	Proper selection of contracting strategy/tender processes
PEM	Active involvement of stakeholders

Declaration of Competing Interest The authors declare that they have no conflict of interest.

Ethical Approval All the procedures concerning the ethical impact of human research performed in this research were reviewed and approved by the Human Research Ethics Committee (HREC) at the Delft University of Technology. All the participants were informed about the goal of the research. Regarding data privacy, the research participants were assured of their anonymity and the case studies before conducting the interviews.

Appendix 1: Sustainability Success Sub-criteria Framework

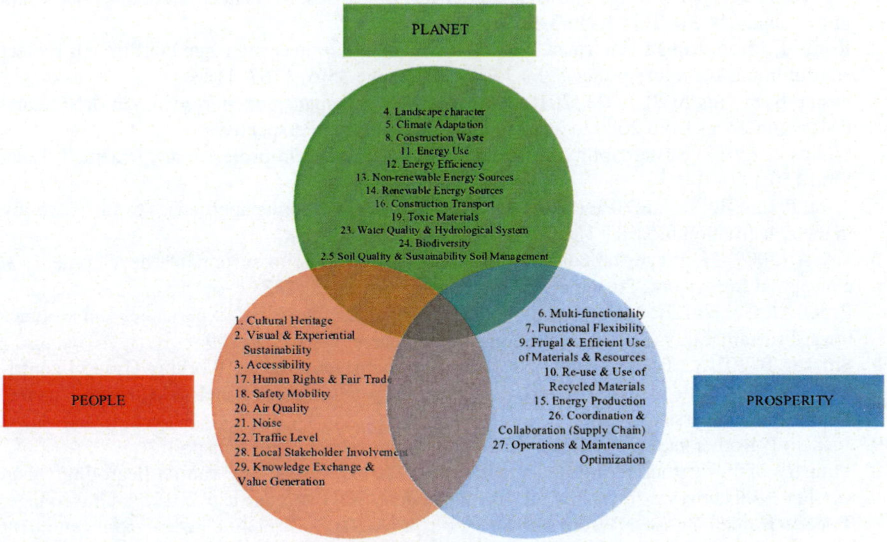

Appendix 2: Selected IPM Roles for the Interview

IPM role in project	Case 1	Case 2	Case 3
Project Manager		×	
Project Control Manager	×	×	×
Project Environment Manager		×	×
Technical Manager	×		×
Contract Manager		×	×
Plan-study Manager	×		

(continued)

(continued)

IPM role in project	Case 1	Case 2	Case 3
Innovation Manager	×		
Total interviewees per case	4	4	4
Total	12		

References

1. Silvius G, Schipper R (2014) Sustainability in project management: a literature review and impact analysis. Soc Bus 4(1):63–96
2. Kivilä J, Martinsuo M, Vuorinen L (2017) Sustainable project management through project control in infrastructure projects. Int J Project Manage 35(6):1167–1183
3. Pasian B, Silvius AGJEAOM (2016) A review of project management research in IRNOP and PMI conferences from 2009 to *2014* to Identify Emerging Perspectives
4. Silvius G (2017) Sustainability as a new school of thought in project management. J Clean Prod 166:1479–1493
5. Afzal F, Lim B, Prasad DJPE (2017) An investigation of corporate approaches to sustainability in the construction industry 180:202–210
6. Sev A (2009) How can the construction industry contribute to sustainable development? A conceptual framework. Sustainable Development 17(3):161–173
7. Pagell M, Gobeli DJP, Management O (2009) How plant managers' experiences and attitudes toward sustainability relate to operational performance, 18(3):278–299
8. Stocker T (2014) Climate change 2013: the physical science basis: Working Group I contribution to the Fifth assessment report of the Intergovernmental Panel on Climate Change. Cambridge University Press
9. Jochem P, Rothengatter W, Schade W (2016) Climate change and transport
10. Ministry of Infrastructure and the Environment (2019) Vaststelling van de begrotingsstaten van het Ministerie van Infrastructuur en Waterstaat (XII) voorhet jaar 2019. 's-Gravenhage: Tweede Kamer. Retrieved from https://www.rijksoverheid.nl/documenten/begrotingen/2018/09/18/xii-infrastructuur-en-waterstaatrijksbegroting-2019. Accessed on 10 Jan 2020
11. Sabini L, Muzio D, Alderman N (2019) 25 years of 'sustainable projects'. What we know and what the literature says. Neil Int J Project Manag 37(6):820–838
12. Tamak R (2017) Project management for sustainability. Delft University of Technology, Delft, The Netherlands
13. de Wit A (1988) Measurement of project success. Int J Project Manage 6(3):164–170
14. Shenhar AJ, Dvir D (2007) Reinventing project management: the diamond approach to successful growth and innovation. Harvard Business Review Press
15. Carvalho MM, Rabechini Jr R (2017) Can project sustainability management impact project success? An empirical study applying a contingent approach. Int J Project Manag 35(6):1120–1132
16. Yuan HJPMJ (2017) Achieving sustainability in railway projects: major stakeholder concerns, 48(5):115–132
17. Elkington J (2013) Enter the triple bottom line. In: The triple bottom line, Routledge, pp 23–38
18. Zimmerman JB (2005) EPA's P3-people, prosperity, and planet-award. Sustain: Sci Prac Policy 1(2):32–33
19. Windhoff-Héritier A, Kerwer D, Knill C, Douillet A-C, Lehmkuhl D, Teutsch M (2001) Differential Europe: the European Union impact on national policymaking. Rowman & Littlefield

20. Gijzel D, Bosch-Rekveldt M, Schraven D, Hertogh M (2020) Integrating sustainability into major infrastructure projects: four perspectives on sustainable tunnel development. Sustainability 12(1):6
21. Mavi RK, Standing C (2018) Critical success factors of sustainable project management in construction: a fuzzy DEMATEL-ANP approach. J Clean Prod 194:751–765
22. Martens ML, Carvalho MM (2017) Key factors of sustainability in project management context: A survey exploring the project managers' perspective. Int J Project Manage 35(6):1084–1102
23. Banihashemi S, Hosseini MR, Golizadeh H, Sankaran S (2017) Critical success factors (CSFs) for integration of sustainability into construction project management practices in developing countries. Int J Project Manage 35(6):1103–1119
24. Slaughter ES (2000) Implementation of construction innovations. Build Res Inform 28(1):2–17
25. Molaei M, Bosch-Rekveldt M, Bakker H (2019) Extending the view on project performance. Adminis Sci 9(3):65
26. Westerveld E (2003) The Project Excellence Model®: linking success criteria and critical success factors. Int J Project Manage 21(6):411–418
27. Hertogh M, Baker S, Staal-Ong PL, Westerveld E (2008) Managing Large Infrastructure Projects, Research on Best Practices and Lessons Learnt in Large Infrastructure Projects in Europe. AT Osborne BV, Baarn
28. Peenstra RT, Silvius AG (2018) Considering sustainability in projects: exploring the perspective of suppliers. IJISPM-Int J Inform Syst Project Manag 6(2):5–22
29. Eisenhardt KM (1989) Building theories from case study research. Acad Manag Rev 14(4):532–550
30. Miles MB, Huberman AM (1994) Qualitative data analysis: an expanded sourcebook. Sage
31. Schramm W (1971) Notes on case studies of Instructional Media Projects
32. Lechler T, Edington B (2013) The silver lining of project uncertainties: discovering opportunities to enhance project value. Paper presented at PMI® Global Congress 2013—North America, New Orleans, LA. Project Management Institute, Newtown Square, PA
33. Silvius G, Schipper R, Visser M (2017) Exploring factors that stimulate project managers to consider sustainability. In: International Research Network on Organizing by Projects (IRNOP), Boston (2017)
34. Hosseini MR, Banihashemi S, Martek I, Golizadeh H, Ghodoosi F (2017) Sustainable delivery of megaprojects in Iran: Integrated model of contextual factors. J Manag Eng 34(2):05017011

Chapter 4
An Alternative Risk Assessment Routine for Decision Making; Towards a VUCA Meter to Assess the Volatility, Uncertainty, Complexity and Ambiguity of Complex Projects

Thordur Vikingur Fridgeirsson, Bara Hlin Kristjansdottir, and Helgi Thor Ingason

Abstract Decisions in today's world face increasingly faster rates of change in complex environments. A prevalent framework for understanding the challenges of complexity in the modern business environment is VUCA. VUCA stands for Volatility, Uncertainty, Complexity and Ambiguity. Research has shown that the unique terms in the VUCA acronym call for different mitigating approaches. In order to mitigate the impact of a VUCA environment on complex projects, organizations and project leaders need to know the type and severity of challenges they are dealing with in each unique project. This research project explores the existing literature and limitations for assessing projects from a VUCA perspective. It aims to draft an accessible diagnostic tool to assess the VUCA dimensions a project is facing, so that project managers can effectively allocate their limited resources. The tool is tested on five complex projects in the manufacturing industry, and its effectiveness is discussed.

Keywords Project preparation · Decisions · Risk management

T. V. Fridgeirsson (✉) · B. H. Kristjansdottir · H. T. Ingason
School of Engineering, Reykjavik University, Menntavegur 1, 101 Reykjavik, Iceland
e-mail: thordurv@ru.is

B. H. Kristjansdottir
e-mail: bara.kristjansdottir@marel.com

H. T. Ingason
e-mail: helgithor@ru.is

R. Cuevas et al. (eds.), *Research on Project, Programme and Portfolio Management*, Lecture Notes in Management and Industrial Engineering,
https://doi.org/10.1007/978-3-030-60139-3_4

4.1 Introduction

The world is changing at a rapid pace. Organizations increasingly turn to projects as a vehicle for transformation and implementation of strategic initiatives to stay competitive. This is generally known as the projectification of the society [1]. Consequently, decision-makers and project managers are increasingly responsible for bringing about successful outcomes in these dynamic environments. However, in spite of increased popularity of projects to make organizations more flexible and innovative projects still are notorious for cost overruns and problems during the implementation see for example, the work of Bent Flyvbjerg, professor at Oxford University that has published an array of works verifying this problem.

It is recognized that flawed decisions in the preparation phase can result in costly and significant problems in the project implementation. Risk management is therefore an important preventive measure when assessing strategic and tactical decision in the project preparation. Traditional risk assessment models generally include evaluating chances of occurrence of a risk event, the impact if incurred and the detection difficulties as numerical values. By multiplication, the coefficient of the risk is established, the highest value presenting the most significant risk event to mitigate, etc. This paper presents an alternative and complementary method to assess the risk. Furthermore, it categorizes the risk in predefined risk groups that indicate the nature of the risk for the project under screening.

The acronym VUCA [2] has become a popular concept in contemporary management science. VUCA stands for volatility, uncertainty, complexity and ambiguity. It is a matrix, as it separates different types of challenges that projects face. With a clear view of which element of VUCA affects a project, and to how great a degree, decision-makers and project managers could more effectively allocate limited resources to achieve successful outcomes in challenging environments (Fig. 4.1).

This study asks whether such a measurement could be achieved through a relatively simple questionnaire applied in the pre-project decision stages. It attempts to determine an effective measurement of each VUCA dimension's impact on a project. The impact of careful risk assessment is greatest at the start, before execution begins, and so decision-makers and project managers benefit from knowing as much as possible during initial analysis on the undercurrents shaping the project. Underestimation of requirements and risk is a common reason for project failure, and in a VUCA world, it becomes a huge challenge to accurately assess requirements and resource needs. A VUCA meter could support proactive project management, serving as a valuable addition to enhance strategic risk awareness, in order to define, analyse and quantify VUCA impact on a project. In this paper, the first draft of such a questionnaire will be put forward and tested on a number of complex projects, to evaluate the accuracy of the measurement.

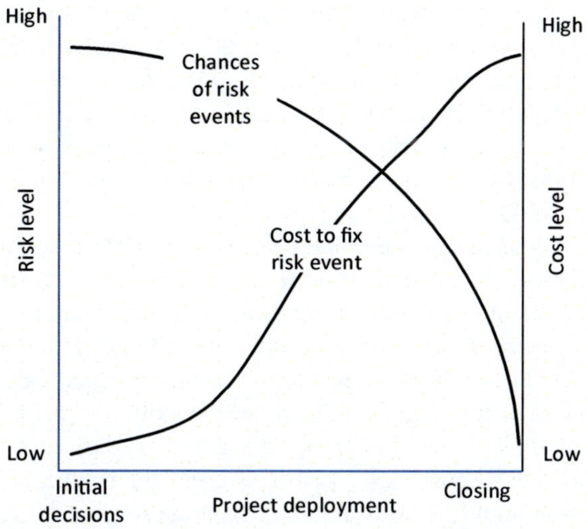

Fig. 4.1 The hypothetical relation between risk event detection and cost to correct the risk event in projects (adapted from Gray and Larson [3])

4.2 Literature Review

Project complexity is a topic in project management research that has received considerable attention. From examining the intersection of complexity theory and project management [4] to mapping specific complexities to specific competencies for project managers [5], project management research has endeavoured to shed light on the best approaches and practices to deal with the ever-changing, and increasingly complex, environment of projects in modern business.

One of the challenges encountered by researchers is the difficulty in defining what, exactly, complexity is. Various approaches and research have attempted to either narrow down the field of application or propose definitions that add dimensions to the understanding of complexity. For example, M. Bosch-Rekveldt et al., propose a framework for the elements that contribute to project complexity in large engineering projects, effectively limiting the field of vision to a specific type of project within a specific industry [6]. In this line of research, VUCA has been proposed as a useful matrix for example, understanding and assessing project environment risk [7].

Another model of project complexity, proposed by Remington, Zolin and Turner, distinguishes dimensions of complexity from severity, in order to address the fact that the: "wide range of factors that may contribute to project complexity […] are in turn subject to a range of severity factors" [8]. By separating the dimensions of complexity and severity, they aim to improve the robustness of the models' measurements. Remington et al's research discusses the existing literature on complexity in some depth, and proceeds to "propose an operational distinction between the terms *dimensions* of complexity, which tells us where the complexity comes from and the *severity*, which tells us to what extent it will be a problem" [8].

VUCA first appeared as an acronym used by the U.S. Army War College. According to the U.S. Army Heritage and Education Center, it was first cited in

Army War College documents in 1987 and became popular in strategic and leadership research throughout the 1990s [2]. The term has made its way into the business lexicon, with popular explanatory articles appearing in leading publications, such as the Harvard Business Review [9] and Forbes [10]. This paper bases its primary definition of the component parts of VUCA for this project on Bennett & Lemoine's article from 2014, "What a difference a word makes: Understanding threats to performance in a VUCA world" [11].

Bennett and Lemoine discuss the growing popularity of VUCA to describe the rapidly changing environment that modern business must navigate. They warn against conflating the distinct terms of the VUCA framework, and of giving up in the face of the seeming unpredictability of VUCA. Despite the myriad popular articles, they claim that "there is a lack of information regarding just what it is that leaders should do in order to confront [...] these conditions" [11]. Properly identifying them, they claim, is crucial to appropriate action, as they each "require their own separate and unique responses. [...] Failure to use the right label will lead to a misallocation of what could be considerable corporate resources" [11]. The authors proceed to lay out a definition of each of the distinct terms [11], summarized in Appendix 1.

This project builds on that analysis in order to differentiate dimensions of VUCA, as it provides a practical application of these definitions. It is important to note that complexity is one of the four concepts that VUCA consists of, but in the previously discussed literature, the VUCA concepts could all be treated as causing "complexity".

4.3 Research Method

In order to prepare the correct project management responses to a VUCA environment, the nature of the risk must first be characterized. The research question posed by this project is whether it is possible to build a "VUCA meter" to gauge the severity of each dimension's impact on a project. The project attempts to measure the severity of each VUCA attribute (strength of V, strength of U, etc.) and produce a numerical value on how VUCA a potential project is in general, as well as the severity of each dimension within the VUCA scope.

The methodology selected for this meter is a questionnaire with closed questions in the form of statements and a Likert rating scale for answers [12]. The Likert scale is a psychometric scale that is commonly involved in research that employs questionnaires. The scale's inventor, Rensis Likert, distinguishes between a scale proper, which emerges from collective responses to a set of items, and the format in which responses are scored along with a range. In terms of separating complexity into *dimension* and *severity,* the terms correspond to the statements and ratings. Dimension of complexity refers to the statement in the questionnaire, as an indicator of where the complexity comes from. Severity of complexity refers to the Likert rating, as a measure of to what extent the dimension will be a problem [8].

Each statement in the proposed questionnaire should apply to any complex project, or activities that should be carried out for any complex project, to a greater or lesser degree, according to the standard of projects as described by the International Project Management Association in their IPMA Standard [13] or the Project Management Institute's PMBOK Standard [14]. Morten Fangel's *Nuanced characterisation of management complexity* [15] questionnaire, was examined for comparison for management complexity evaluation questionnaires. In addition, the content of the statements was compared to the risk factors analysed in the literature review above.

The rating scale measures the severity of each subject's estimation of complexity in the particular dimension of the project or project activities indicated by the statement. A five-point range was selected for this project. In order to maintain the questionnaire's accuracy, number of questions per VUCA attribute are the same. It is assumed that the evaluated statements have high internal consistency and so a high score answer to one question has the same weight on a total score as a high score answer to any other question (interval range). This is a limitation of Likert scales in general. Trends in weights across categories and different subjects will be examined in the results to test that assumption.

Closed statement questionnaires with rating scales are a form of qualitative research that delivers quantitative results based on subjective evaluation. Each answer is based on the evaluation of the subject, which can be influenced by that individual's perceptions, experiences or emotions. The resulting data takes numerical form, which can provide a measure of consistency and measurability. This makes the results easier to evaluate and provides an added benefit in use, e.g. for an organization that wishes to see trends over time.

The questionnaire is set up in an electronic form (using Microsoft Forms) and sent to the selected project managers. Excel was used to analyse the results. The subjects belong to one globally operating organization, with established technical risk-level categories for engineering projects. All but one of the projects are of the highest risk level. One project involves a complex IT implementation. The projects' managers give answers to the questionnaire without seeing the evaluated score. A more effective test might be to apply the questionnaire before a project starts and preparing a comparison after its completion, but due to the time constraint of this study, the best attempt is made within its limitations in a single point in time.

Building on the definitions by Bennett and Lemoine, key components of each distinct VUCA term are captured. Statements are formed with regard to typical questions in previously discussed risk assessment questionnaires. The statements are grouped and scored based on the relation to the type of risk associated with a given VUCA term. The resulting questionnaire takes the form of five statements per VUCA concept, each with a five-point answer range with a numerical weight as interval variables. The answers are: Strongly agree (1), Agree (2), Neither agree nor disagree (3), Disagree (4) and Strongly disagree (5). An interval scale such as this has ordered numbers with meaningful divisions. In order to give a result that can be averaged, the relative weight of the ratings and statements needs to be as equal as possible. The answer scores are then summed and averaged for each group, Volatility, Uncertainty, Complexity and Ambiguity, as well as a total average score for how VUCA the project maybe, is calculated.

The outcomes are compared and discussed with the subjects, to evaluate whether the meter presents an accurate or inaccurate outcome. Each evaluation takes place in an individual interview. Evaluation of results is a combination of subjective evaluation (categorizing accuracy as high, medium or low) with notes for improvement, and comparison to project actuals (time, cost, etc.). As evaluations are partly subjective, the average score accuracy should indicate whether the intervals and statement categories are correct. Based on the evaluation, a strength rating of High, Medium or Low, is given to each category average result. Based on those evaluations, the meter is adjusted. The first version of a VUCA meter is then presented for future testing and conclusions drawn as to whether it can serve as a rough indicator of how much and wherein the VUCA matrix a project is situated.

The VUCA concepts as described by Bennett and Lemoine have been summarized as follows in order to group them in statements:

Volatility

Unstable and unpredictable resource cost and/or availability at unpredictable times and durations
Expected fluctuations on resources with unknown timing and magnitude

Questions aim to capture key reasons why resources might bc suddenly unavailable or expensive and the challenges in resolving resourcing for unforeseen new needs.

Uncertainty

Lack of knowledge
Unclear impact of change, but cause and effect are known

Questions aim to capture key risks for lack of comprehensive and reliable knowledge in projects, or difficulty in communicating or accessing relevant information or effectively resolving questions on unforeseen new issues.

Complexity

Many interconnected parts
Complex regulatory/political environments, multiple component parts

Questions aim to capture key risks for the structural or systems-related complexity in the environment, internal and external, of a project, as well as whether a governance to effectively deal with these complexities is in place.

Ambiguity

Doubt about the nature of cause and effect
Little to no historical information to predict the outcome
Difficult to forecast or plan for

Questions aim to capture whether there are risks of lack of experience and predictability that might affect the project, and whether the project delivers unpredictable new changes, either internal or external to the project and organization.

4.4 Research Results

Average answer time for the questionnaire was 11 min and 40 s. Five subjects were selected by a globally operating manufacturing organization, all project managers for complex projects.

The projects examined with the meter are

1. Complex engineering project in collaboration with multiple other organizations/technologies
2. New ERP implementation project
3. Complex engineering project
4. Complex engineering project
5. Complex engineering project

Volatility

The result of Volatility scores indicates that the meter gave a highly accurate result.

Project #	VUCA score	Accuracy evaluation
1	2.8	High
2	3.2	High
3	4.4	High
4	2.8	High
5	3.6	High

Subjects #1 and #2 indicated that the volatility was accurate but in reviewing the results they might be slightly lower. However, both subjects also stated that it could be lower due to the experience and culture that facilitates resource slack (mitigating approach), not because the measurement of complexity was inaccurate. Statement phrasing might be re-evaluated based on this so that indications of existing practices to mitigate the risk are reflected in the rating.

Uncertainty

The result of Uncertainty scores indicates that the meter gave a highly accurate result.

Project #	VUCA score	Accuracy
1	3.8	High
2	3.2	High
3	4	High
4	3.4	High
5	4	High

In interviews, subjects #3 and #4 both emphasized the impact of Uncertainty on the other risk factors in the project. The impact of Uncertainty on Volatility was discussed, as a lack of reliable information in the projects had a direct impact on resource needs, which in turn magnifies the impact of volatility in resources.

Complexity

The result of Complexity scores indicates that the meter gave a fairly accurate result.

Project #	VUCA score	Accuracy
1	4.4	High
2	3.8	High
3	3	Medium
4	2.4	High
5	4.2	High

Subject #3 stated that the score for complexity may have needed to be higher, although its relative value compared to other categories was accurate. In discussing the complexity and scores, the complexity factor of organizational departments could be better phrased to indicate the organizational complexity of globally operating companies. This was a site of some complexity, though the VUCA impact was most felt in the Uncertainty dimension, as the complexity led to unreliable information assumptions. In addition, phrasing could be adjusted to also represent projects involving multiple companies where one organization is leading the project management.

Ambiguity

The result of Ambiguity scores indicates that the meter gave a highly accurate result.

Project #	VUCA score	Accuracy
1	3	High
2	2.6	High
3	3.6	High
4	2.6	High
5	4	High

Subjects #1, #2 and #4 indicated that the ambiguity score, while lower than might have been expected given the nature of the projects, gave an accurate depiction of the ambiguity reflected by project actuals (time, cost, progress). Both subjects stated that strong application and experience of project management methodologies and clear structure of the projects were the reason for this.

Total VUCA Score

The total VUCA Score as calculated was as follows:

Project #	VUCA score
1	3.5
2	3.2
3	3.75
4	2.8
5	3.95

Subjects did not see much value in a single numerical score but stated that the overall graphical charts showing the relative severity of dimensions within the projects were useful and gave a highly accurate map of the project challenges.

Individual answer score ranges

Each VUCA category scored on a range of 2.x–4.x across the different projects. Answer ranges for individual statements were as follows (Table 4.1).

Three statements have a very low range of scoring across all projects. The statements involve sequential planning, distributed stakeholders and recognition of stakeholder relationships. Uniformity in scores on stakeholders across cultures and time zones may be explained by the fact that the organization in question is globally operating and few complex projects within the organization involve only one location. Sequential planning scores may similarly be explained with the nature of the projects under investigation. The final statement in the questionnaire has no strong answer scores and could be refined to further draw out the potential complexity of the ambiguity dimension. Further testing should be conducted to examine if the range remains uniform in other organizations that deal with complex projects.

4.5 Discussion

Interviews with all subjects gave strong confidence to the overall representation of where projects were situated in the VUCA matrix. The graphical representations of a severity score for each dimension were described as the most useful result. The comparative severity of each dimension became clear in the graphs and was rated as highly accurate by each subject.

Overall, trends in weights indicate high internal consistency of statement scores. Comparing scores across projects indicates that the subjective evaluation influences the numerical score. Numerical scores without context were described as less helpful. While the intervals between ratings were accurate, the tool is limited by the fact that evaluation is subjective and the strength of numerical values is thus influenced by each subject. For example, a subject may be prone to avoid extremes of the answering scale, and so the subjective nature of evaluation can lead to higher/lower numerical scores for all answers, but as all scores are given by the same person, the relationship between the severity of dimensions remains accurate. The individual numerical scores are therefore less useful, and the value was seen in the relative

Table 4.1 Compliance to VUCA criteria's

Volatility	#1	#2	#3	#4	#5
Simple in planning (straightforward/sequential execution)	5	5	5	4	5
Resource needs are known and accessible	2	4	4	4	2
Adequate timeframe with good slack in schedule	3	4	4	2	4
Solid contracts throughout project duration	2	2	5	2	5
Known, well-defined objectives	2	1	4	2	2
Uncertainty					
Uses few and proven technology components	5	4	5	4	2
Stakeholders are few, with few time zones/cultural differences	5	5	5	4	5
Information is easy to obtain	3	4	4	4	5
Scope is well defined and approved	2	1	2	2	4
Risk management is well defined	4	2	4	3	4
Complexity					
Few and simple regulatory or political environments	5	5	2	2	4
Few subcontractors, organizational departments, and cultural differences	5	5	2	2	5
Few interfaces with other technologies, projects or operations	5	5	5	2	4
Has been done many times before	4	3	2	4	4
Clear governance, straightforward decision-making	3	1	4	2	4
Ambiguity					
Deliverables are well defined, no "unknowns unknowns"	5	2	4	3	5
Connections between tasks are clear	2	4	3	2	2
Risk factors are well known and documented	4	2	4	4	5
No "hidden agenda"	2	3	4	2	5
All stakeholders and their relationships are recognized	2	2	3	2	3

difference between VUCA dimensions. The total average score is therefore also not of particular value in this measurement approach.

This means that a graphical view of all category severity scores gave a good representation of the projects' key challenges from a VUCA perspective for all subjects. The results are useful as input for risk assessment, requirement gathering, project scoping and approach, and to communicate the nature of project challenges to stakeholders. Visual representations can be very helpful in communicating the project situation to organization leadership. For example, the results of the VUCA scores can be displayed as a graphical chart such as Fig. 4.2, or as radar maps (Fig. 4.3).

With a chart such as the above, the VUCA landscape of the project becomes clearer, and appropriate measures can be taken to address the unique project situation.

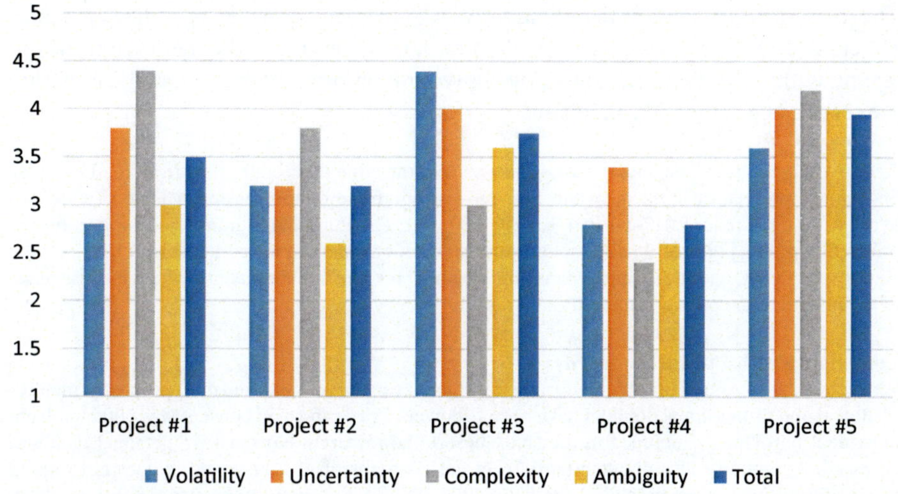

Fig. 4.2 Results overview

Fig. 4.3 Example of VUCA radar map

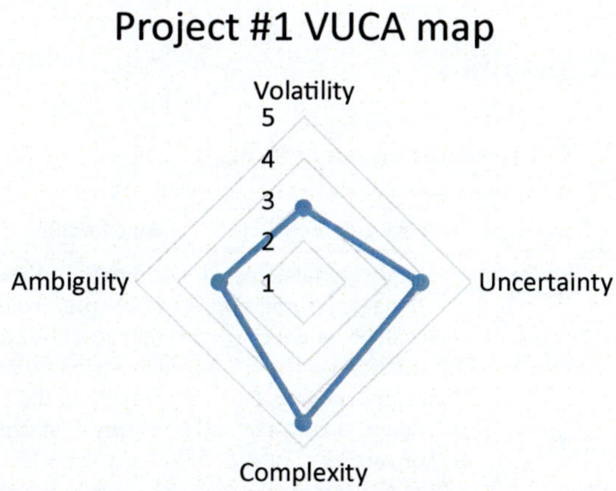

4.6 Conclusions

The literature research shows that complexity and risk need to be viewed from multiple angles. The results of the VUCA meter tests illustrate the importance of clarifying the particular project situation and dimensions of complexity that projects face. While an aggregate risk score can indicate a generic need for a project management response, the VUCA meter as developed by this project was evaluated as most useful for evoking discussion and risk awareness, with a graphical comparison of the relative severity of different types of risk. In complex environments with limited resources, this aids in prioritization that could significantly influence project success.

The literature on responses to individual VUCA elements can then be utilized to better effect. The meter presented here is the first attempt and would benefit from further testing within a longer timeframe, by applying it before project start, and comparing the results at a project's conclusion.

Acknowledgements The authors would like to acknowledge their colleagues from the Decision-Ship Ahoy project, co-funded by the Erasmus+ programme of the European Union (www.dahoyproject.eu, reference 2017-1-FR01-KA203-037301). The European Commission support for the production of this publication does not constitute an endorsement of the contents which reflects the author views only, and the Commission cannot be held responsible for any use which may be made of the information contained therein.

Compliance with Ethical Standards
We declare that I ensured the objectivity and transparency in our research and that accepted principles of ethical and professional conduct have been followed. Prior informed consent was obtained from individual participants included in the study before the research. No sensitive personal data was accessed. Anonymity of individual participant data is maintained. The research does not require ethics approval, as is mentioned in the waiver no. 09-47-28, issued by the Reykjavík University (RU) Ethics Committee.

Appendix 1

VUCA concepts and definitions.

	What it is	An example	Recommended response
Volatility	Relatively unstable changes; information is available and the situation is understandable, but change is frequent and sometimes unpredictable	Commodity pricing is often quite volatile; jet fuel costs, for instance, have been quite volatile in the twenty-first century	Agility is key to coping with volatility. Resources should be aggressively directed toward building slack and creating the potential for future flexibility
Uncertainty	A lack of knowledge as to whether an event will have meaningful ramifications; cause and effect are understood, but it is unknown if an event will create significant change	Anti-terrorism initiatives are generally plagued with uncertainty; we understand many causes of terrorism but not exactly when and how they could spur attacks	Information is critical to reducing uncertainty. Firms should move beyond existing information sources to both gather new data and consider it from new perspectives

(continued)

(continued)

	What it is	An example	Recommended response
Complexity	Many interconnected parts, forming elaborate networks of information and procedures; often multiform and convoluted but not necessarily involving change	Moving into foreign markets is frequently complex; doing business in new countries often involves navigating a complex web of tariffs, laws, regulations and logistics issues	Restructuring internal company operations to match external complexity is an effective and efficient way to address. Firms should attempt to 'match' their own operations and processes to mirror environmental complexities
Ambiguity	Lack of knowledge as to the 'basic rules of the game'; Cause and effect are not understood and there is no precedent for making predictions as to what to expect	Transition from print to digital media has been very ambiguous; companies are still learning how customers will access and experience data and entertainment given new technologies	Experimentation is necessary for reducing ambiguity. Only through intelligent experimentation can firm leaders determine what strategies are and are not beneficial in situations where former rules of business no longer apply

Definitions laid out by the authors based on [11]

References

1. Schoper Y-G, Wald A, Ingason HT et al (2018) Projectification in Western economies: a comparative study of Germany, Norway and Iceland. Int J Proj Manage 36(1):71–82. https://doi.org/10.1016/j.ijproman.2017.07.008
2. U.S. Army Heritage and Education Center (2018) USAHEC ask us a question. (2018, February 16) Retrieved from U.S. Army Heritage and Education Center. http://usawc.libanswers.com/faq/84869
3. Larson E, Gray C (2018) Project management: the managerial process, 7ed. McGraw-Hill Education, New York, NY, p 10121
4. Cooke-Davies T, Cicmil S, Crawford L, Richardson K (2007) We're not in Kansas anymore, Toto: mapping the strange landscape of complexity theory, and its relationship to project management. Proj Manage J 38(2):50–61
5. Bosch-Rekveldt MM (2009) Mapping project manager's competences to project complexity. In: Kakonen K (ed) IPMA 23rd WorldCongress, research track human side of projects in modern business. Project Management Association Finland (PMAF) and VTT Technical Research Centre of Finland, Helsinki
6. Bosch-Rekveldt MJ (2011) Grasping project complexity in large engineering projects: the TOE (Technical, Organizational and Environmental) framework. Int J Proj Manage 29(6):728–739
7. Szpitter A, Sadkowska J (2016) Using VUCA matrix for the assessment of project environment risk. Zarzadzanie i Finanse 401–413
8. Remington K, Zolin R, Turner R (2009) A model of project complexity: distinguishing dimensions of complexity from severity. In: Proceedings of the 9th international research network of project management conference, 11–13 Oct 2009, Berlin
9. Bennett N, Lemoine GJ (2014) What VUCA really means for you. Retrieved from Harvard Business Review: https://hbr.org/2014/01/what-vuca-really-means-for-you

10. Kraaijenbrink J (2018) What does VUCA really mean? Retrieved from Forbes: https://www.forbes.com/sites/jeroenkraaijenbrink/2018/12/19/what-does-vuca-really-mean/#43e7ed4d17d6
11. Bennett N, Lemoine GJ (2014) What a difference a word makes: understanding threats to performance in a VUCA World. Bus Horiz 57(3):311–317. https://doi.org/10.2139/ssrn.2406676
12. Allen E, Seaman C (2007) Likert scales and data analyses. Qual Prog 40:64–65
13. IPMA. IPMA standard. Retrieved from IPMA.com: https://www.ipma.world/projects/standard/
14. Project Management Institute. Foundational standards. Retrieved from PMI.org: https://www.pmi.org/pmbok-guide-standards/foundational
15. Fangel M (2013) Nuanced characterization of management complexity. In: Fangel M (ed) Proactive project management. Fanel Consulting AS, Hilleroed, Denmark, pp 38–39

Chapter 5
Sustainable Project-Oriented Careers: A Conceptual Model

Gregory J. Skulmoski, Craig Langston, Alan Patching, and Amir Ghanbaripour

Abstract An area of interest for project participants is career management. Some active network to find their next consulting engagement, upskill or prepare for major life changes (e.g. marriage or promotion). We find in the literature improved results for those who pursue a sustainable career path. We examine sustainable project-oriented careers and offer error messages that indicate career turbulence such as infrequent upskilling. The sustainable project-oriented career model is reviewed within the context of the Gig Economy where significant turbulence exists (e.g. COVID-19 pandemic) with the mediating role of training and continuing education. We conclude with an overview of innovations in post secondary education that facilitate a sustainable project-oriented career. We bring together evolving career sustainability concepts within the project environment and provide guidance for developing a sustainable project-oriented career, especially important in our current tempestuous times.

Keywords Sustainable project-oriented career · Systematic literature review · Sustainable career turbulence · Sustainable career drivers

---

G. J. Skulmoski (✉) · C. Langston · A. Patching · A. Ghanbaripour
Bond University, 14 University Drive, Robina, QLD 4226, Australia
e-mail: gskulmos@bond.edu.au

C. Langston
e-mail: clangsto@bond.edu.au

A. Patching
e-mail: apatchin@bond.edu.au

A. Ghanbaripour
e-mail: aghanbar@bond.edu.au

R. Cuevas et al. (eds.), *Research on Project, Programme and Portfolio Management*, Lecture Notes in Management and Industrial Engineering,
https://doi.org/10.1007/978-3-030-60139-3_5

5.1 The Classic Project-Oriented Career

Modern project management is a relatively new phenomenon despite the long history of projects throughout millennia; we did not use modern terms to describe what was happening (e.g. even though we had a project sponsor for building the pyramids, they were not called project sponsors; they were simply called pharaohs). It was not until the twentieth century that modern project management emerged as we know it today. Carayannis et al. [12] reviewed the history of project management and traced the shift of focus from tools (e.g., PERT and Monte Carlo) to the human element (e.g. matrix organization and human resource management). We see professional organizations like the International Project Management Association form, grow and offer certifications to foster project management competence and the project-oriented career.

During this time, organizations also long relied on projects to bring new products and services to customers. Favaro [24] distils the history of business strategy, moving from efficiency targets, through to increasing scale and improving quality, to the rise of integrated networks and ecosystems to serve customers. To achieve the strategy, organizations would plan for 5–10 years and then launch a series of carefully and fully planned projects. Many organizations transitioned to a project-oriented organization to help achieve their strategy [28].

5.1.1 Accidental Project Manager

The project-oriented career began to evolve in this time of change. In the past, a typical career in projects began when a person joins a project team as a junior member. If they do well, then they might be promoted to a "senior" project participant (e.g. Lead Engineer). Then if they succeed, we stop them from what they do well and then give them the new role of project manager often without formal training (or they attend project scheduling software training). There is little wonder looking back that we have not equipped our project teams to succeed since many "accidental" project managers led these challenging projects [32].

Professional organizations offer project-oriented certifications to address competency gaps:

International Project Management Association (IPMA)—Certified Project Manager,
Project Management Institute (PMI)—Project Management Professional,
Global Association for Quality Management (GAQM)—Certified Project Director,
British Computer Society (BCS)—Chartered Professional,
Association for the Advancement of Cost Engineering International (AACEI)—Certified Cost Professional,
EC Council—Project Management in IT Security.

Therefore, we see that the project-oriented career has become more formalized, requiring training and education in areas in addition to technical competence like engineering and project management tool proficiency such as budgets and schedules (Hartman 1999). Planning a career in projects has become much more complex due to significant changes occurring in the business world. In the past, senior management would identify strategic goals, then launch a series of carefully planned projects; however, the timelines now are much more compressed, and the business environment is anything but stable. Agility, disruption, data analytics and integration are the new business mantra [24].

5.2 The Gig Economy and Other Systemic Changes

Long-term planning and a stable workforce were characteristics of the business environment that have long since passed. Now, we have a Gig Economy at the front end of the 4th Industrial Revolution. Indeed, Akkermans, Seibert and Mol [3] state that a predictable career and lifetime employment with stable firms is a thing of the past. In this changing time, project-oriented people may re-examine their career path to be contributors rather than sidelined.

The Gig Economy may simply be described as temporary employment where contractors contribute to an organization's efforts through hourly or piecemeal work. For example, an independent contractor may join a project to set up and test a computer server, then leave the project when their contracted work is completed. Or the Gig Economy worker may be paid on a task-by-task basis such as delivery drivers (e.g. Uber drivers). Organizations now have more opportunities to hire temporary workers when required and minimize a core structure of full-time/long-term employees [46].

Organizations increasingly are hiring temporary workers when needed, some in-house services are contracted to external contractors to improve organizational effectiveness and efficiency, by having non-core activities performed by contractors to allow an organization to focus on core activities. However, some organizations choose the outsourcing model to bring in highly skilled contractors to bring innovation to core areas of business [23]. The global demand for outsourcing in 2015 was $524.4 billion and rose to $565 billion 2 years later. This growth has been primarily in logistics, IT (information technology) and business processing such as financial, human resource and call centres [40]. Indeed, IT outsourcing on a global level is expected to grow; such as in 2017 when outsourcing grew by 5.7% in that single year [23]. Indeed, some industries prefer the outsourced model rather than relying on in-house services: by 2020, 72% of all global clinical trials of pharmaceutical products is expected to be conducted by contract organizations with an annual growth rate of 6.9 annually [85].

5.2.1 The 4th Industrial Revolution: Challenges and Opportunities

At the same time that the Gig Economy offers increased employment options, and business is more favourable to outsourcing, technology advances promise great change, opportunities and challenges. The 4th Industrial Revolution (also known as Industry 4.0) is the continuation of step changes to business beginning with the 1st Industrial Revolution (see Fig. 5.1). We have progressed by applying steam, electricity and basic information and computer technologies to production. Now we see disruptive technologies being applied to virtual and physical activities on a global scale to allow organizations to cooperate to deliver new products and customizations in ways never before imagined possible or even demanded by consumers [5].

Disruptive technologies are those resources or tools used to produce things (e.g. artificial intelligence) or a finished product with disruptive characteristics (e.g. self-driving car). Other disruptive technologies include blockchain, gene sequencing, large-scale energy storage, building information modelling, nanotechnology, biotechnology, quantum computing, robotics, the Internet of Things and other transformative technologies that can have dual potentialities to help and/or harm [69]. For example, we may see ingestible robots that repair injuries from within, and we may see jobs displaced by these disruptive technologies. Indeed, a recent survey of 1730 project participants reported that 91% of them were impacted by disruptive technologies [64]. Indeed, there are technical "pessimists" who predict severe net job loss due to these disruptive technologies with significant impacts on society [69]. How is business impacted beyond the common challenges of global competition, shorter production cycles and consumer demand for customization? What is the impact to project participants?

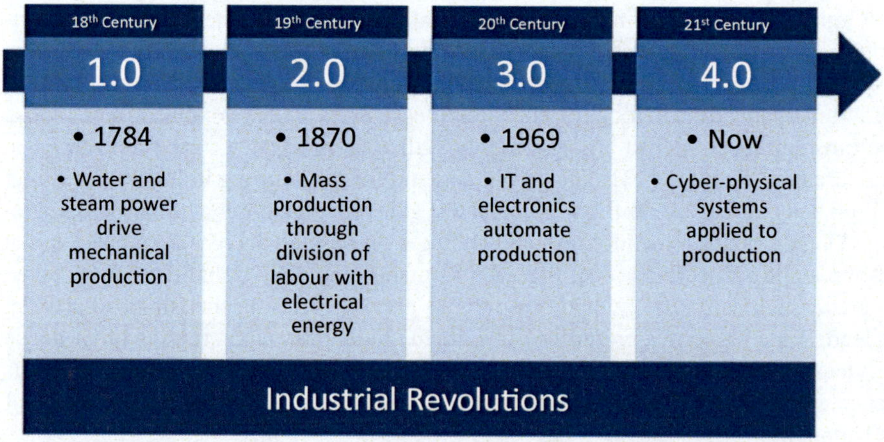

Fig. 5.1 4th industrial revolutions (adapted from 123RF, 2019)

The 4th Industrial Revolution will likely impact the business by changing consumer expectations (e.g. sustainability), product enhancement (e.g. add digital capabilities to products), collaborative innovation (e.g. work with others) and the way organizations form and organize [69]. The news media is ripe with businesses failing thanks to disruptive technologies: we know about the drastic decline of Kodak, Blackberry and Blockbuster. Technology laggards are more likely to lack a mature digital business strategy, are risk-averse, and technology adoption is a low priority [64]. Many organizations may struggle against competitors who are agile and innovative, and harness global digital platforms for research, development, marketing, sales and distribution to outperform digital laggards [69]. Unfortunately, even if organizations have a digital strategy to deliver innovation, approximately one in ten organizations fail to deliver that strategy [64]. That is, even if an organization has a well-thought-out strategy delivered through innovation, they are likely to struggle to deliver that strategy through a project approach. But it is not organizations that deliver—it is people who deliver innovations through projects to achieve the strategy. Therefore, [27] recommends skill renewal to remain competitive.

5.2.2 New Project-Oriented Opportunities

Within these changes—Gig Economy, globalization, outsourcing, disruptive technologies, 4th industrial revolution and more—project-oriented workers who desire to succeed, may think strategically about their careers. What does the project-oriented work forecast look like? Are there jobs in project management despite job loss predictions associated with the 4th Industrial Revolution? Crawford, French and Lloyd-Walker [18] state that "*a growing percentage of the Australian workforce is employed in project-based or project-oriented organisations, leading to the claim that Australia is a project-based economy*". A recent study found that global organizations (chiefly in the financial services, industrial and manufacturing, consumer services, government and non-profit and retail and hospitality sectors) involved in delivering innovative projects rank the technical talent shortage as their top risk to strategy success [27]. Indeed, this shortfall is validated in a PMI study [66] that also looked at the supply and demand of project-oriented people and forecasted a talent shortage:

By 2027, organizations will need 87.7 million people in project-oriented roles;
The talent gap could result in a loss of $207.9 billion in GDP through 2027;
On an annual basis, organizations will need to fill 2.2 million project-oriented jobs each year through to 2027.

Thus, while there could be an overall net job loss, there appear to be opportunities in technical areas, and especially in project management for people with the right skills. Technical proficiency ("Project Management Technical Quotient"—PMTQ) with disruptive technologies is an emerging project-oriented skillset: the ability to select, modify, manage and integrate technologies to solve problems and add value to specific projects [63]. Those with PMTQ competencies have three distinctive characteristics in addition to long-standing competencies like soft skills, project management tools and process expertise in traditional and adaptive delivery methods like Agile, and business strategy skills [63]:

Curiosity: they welcome new ideas and ways of doing things wherever it may lie on the low to high technology continuum. They have an open mind, tempered with some scepticism.
Inclusive Leadership: they look after not only their team but their team's resources such as technology and robots.
Future-Proof Talent Pool: they recruit and nurture those who embrace these digital skills, and keep up with trends and adapt their own skill sets accordingly; there is a regular renewal of skills.

When a project-oriented person has these updated skills, including PMTQ, then they are more likely to find and keep jobs in project-oriented organizations [65] or to deliver value through an outsourcing arrangement. This raises the question: what can be found in the project-oriented literature and research?

5.3 Sustainable Project-Oriented Career

To learn more about sustainable careers, we follow a systematic literature review process that involves two steps: plan the literature review process, then follow the literature review plan [11]. We conduct a Boolean search of key career sustainable terms (see Table 5.1) and you searched across over 170 databases, including the following electronic databases such as ISI Web of Science, Scopus and SpringerLink. After searching the entire databases (e.g. everywhere), we narrow our Boolean search to filter in those articles from peer-reviewed journals. We then search only in the title of the publications of peer-reviewed journals.

We follow [11] systematic literature review process and applied inclusion and exclusion criteria (see Table 5.2) to guide the quality of our review. One might be surprised that only 12 journal papers were found with the term "sustainable career" in their titles.

We then read the publications identified in Step 2, Table 5.1.

At some point in our careers, we were hired or brought onto a project team; we had the necessary competencies to add value. However, with time comes unpredictability,

Table 5.1 Search terms

Search actions	Results returned
1. Search "sustainable career" in everywhere	589
2. Filter for results from only peer-reviewed journals	302
3. Search "sustainable career" in title	12

Table 5.2 Inclusion and exclusion criteria

Criteria	
Inclusion criteria	Publications that are related to career sustainability Publications that are related to elements of the Sustainable Project-Oriented Career Publications that were peer-reviewed
Exclusion criteria	Publications that are did not have full-text available Publications that were not in English Publications that incidentally referred to elements of the Sustainable Project-Oriented Career

change and new technology; not all project participants have kept up with advances and are at risk of not being selected for future projects if they do not have the right skills (e.g. their PMTQ is weak). What has to happen for career continuity? Is there such a thing as a sustainable project-oriented career (SPOC)?

Sustainable careers are becoming an increasingly pronounced concern for many as evidenced in the recent growth of research papers in scientific journals and conferences [21]. Indeed, even the meaning of career is in flux as we see more workers adopt a subjective view of career success rather the more traditional view of career success through hierarchical progression [20, 79]. One emerging career sustainability model incorporates agency theory from psychology (the individual obeys the organization in exchange for the organization taking responsibility for their actions); it brings together the dimensions and indicators of career sustainability [21]. The DeVos career sustainability model begins with the concept of time in that sustainability is related to what happens over one's career to protect and foster human and career development. We have modified the model to substitute a project-oriented career occurring over time. It is over time that we can better assess sustainability rather than during a discrete moment of analysis. We have modified this model to specifically explain a project-oriented approach, rather than a psychological approach to career sustainability.

The key modification to this career sustainability model (see Fig. 5.2) is to replace the individual's agency modelling perspective with a person in a project-oriented career, working overtime on a series of projects. In project-oriented organizations, a person may temporarily join an organization to perform project work, and then leave the organization upon completion of that work. However, some join an organization, and then stay for a series of projects, and then leave the organization [38]. It is this concept of a person working in a series of projects that displaces psychological concepts related to career sustainability.

When one has a project-oriented career, and one works in a series of projects, the notion of career sustainability may eventually be considered. Career sustainability is indicated by three elements:

(i) *Productivity*, where the individual performs successfully in their current role as well as having a high potential for future employability. New joiners and existing workers may need to learn new skills to complete tasks.

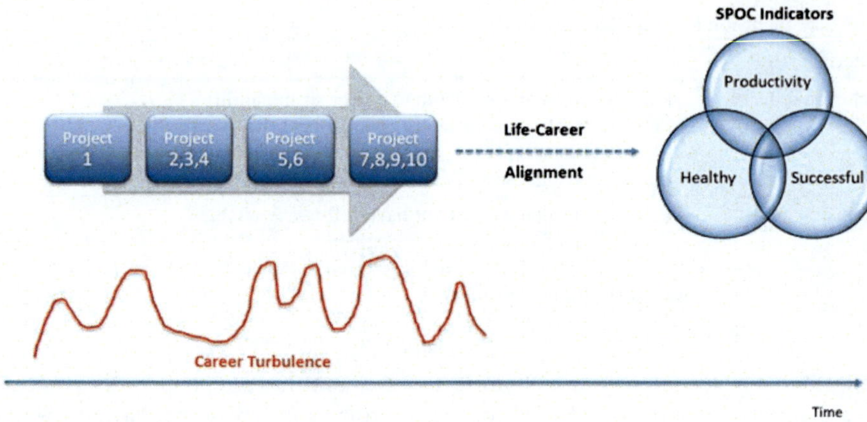

Fig. 5.2 Sustainable Project-oriented career

(ii) *Health*, where a holistic and comprehensive view of health fits with the demands of one's career. With time, health can change: workers begin jobs and initially meet the demands of a job, but with age, one may be less capable (e.g. a job in construction where it becomes more difficult with age to complete tasks requiring manual labour).

(iii) *Happiness*, where the subjective elements relate to holistic satisfaction and feeling successful in one's career—past and present, and one's life outside the career occur.

An important concept is a fluid fit between these three career sustainability indicators: all three may be in flux and impact each other over time [21]. Thus, a project participant works in a series of projects over their career, and if the person has the right mix of success, health and productivity over time, that person has had a sustainable career.

5.3.1 Sustainable Career Turbulence

During one's career, there is the notion that chance events (positive or negative) can occur that alter career trajectory; Akkermans, Seibert and Mol [3] refer to this as career shocks. The worker has control over some shocks like a decision to get pregnant that can impact their career. However, there are some shocks that the worker has less control such as when a person is made redundant or promoted to a position with more responsibility requiring more effort. These career shocks vary in intensity (moderate to severe), valence (positive or negative effects), frequency and duration [21]. These career shocks may also be predictable or unpredictable, as well as originate from multiple sources such as geopolitical, environmental, organizational, interpersonal and family-oriented [3]. Given enough time, most working people face a career shock; how a person reacts or indeed, plans for the career shock, impacts career sustainability. In this paper, we use the term "turbulence" rather than shock, since we can often avoid the ill effects of turbulence, whereas shocks imply less control. Therefore, we substitute career turbulence for career shocks in our model. In Table 5.3, we identify examples of career turbulence from our systematic review of the literature.

Thus, the Sustainable Project-Oriented Career model has the project participant working in a project-oriented organization either as a contractor or as an internal employee. This person works on a series of projects over time, and this is called a project-oriented career. During this career, there are most likely career shocks (positive and negative, that vary in intensity) that face the project participant. If the person makes the right decisions, then there will be a fit between the career and the career sustainability indicators of being successful, healthy and productive will be positive. If the person does not respond appropriately to career shocks, then a sustainable project-oriented career is in jeopardy. Increasingly, a self-directed or "protean career" is the responsibility of the individual rather than the organization [3].

When we reviewed general literature for elements of turbulence, we find potential examples of career turbulence that could cause challenges for project-oriented people:

- The globalization of work and changes to how work is structured,
- Organizations reducing the amount of responsibility for workers' careers,
- Career planning and development procrastination,
- Increasing personal competency and threshold skill deficits, and
- Increasing stress and a work-life imbalance.

The literature also reveals things people should do who have sustainable careers; we refer to these enablers as sustainable career drivers.

5.3.2 Sustainable Career Drivers

There are three key areas of response a person can take to improve the probability they have a sustainable project-oriented career. Repeatedly, the notion of career renewal is highlighted as critical to a sustainable career [1, 50, 51, 78]. It has been long recognized that individuals might complete formal training or education in areas of demand (especially future demand) so that one's skills align with the needs of the organization [72]. Second, one's career is more likely to become sustainable if one is flexible and adaptable to new opportunities; simply, the more one can do, the more opportunities there will be to choose, resulting in career continuity [51, 78]. Finally, building upon work-life balance research, career sustainability is more likely if the career is integrated with one's home life, the community and society [1, 58, 59, 78, 86]. We reviewed the literature for elements related to our Sustainable Project-Oriented Career model and uncovered potentially applicable sustainable career drivers (see Table 5.4).

Table 5.3 Turbulence in sustainable careers

Author	Year	Empirical	Focus	Turbulence
Iles [41]	1997	Opinion	Sustainable career development of high-potential employees	Acceleration of organizational restructuring, downsizing, outsourcing and delayering
Stronach [74]	2010	$N = 14$, mixed	Career sustainability of Indigenous Australian boxers	Lack of mentoring
Asuquo [4]	2013	$N = 210$, surveys	How to build a sustainable career in environmental anthropology	Not staying connected with your community of practice
De Vos [20]	2017	Opinion	Career sustainability in general	Longer careers due to delayed retirement, shorter and less predictable sequence of jobs, more alternative modes of work (e.g. part-time, contracting, etc.), fading boundaries between work and non-work, new ways of working, personal career development accountability and rise of subjective career success
Van Driel [80]	2017	$N = 18$, interviews	Retaining Australia General Practitioner research leaders	Switching institutions, working in isolation, work-life balance, research as a hobby, lack of professional standing, managing dual careers, lack of research funding, low remuneration
Fournier [25]	2018	$N = 61$, interviews	Job loss of older Canadian workers	Globalization, increasing competition, more technology, unstable ties to work
McDonald [51]	2018	Opinion	Creating sustainable careers	Automation, reduction of permanent jobs and a shift to temporary, contract and other forms of work, advanced technology, perceived work insecurity, job precariousness

(continued)

Table 5.3 (continued)

Author	Year	Empirical	Focus	Turbulence
Venkatraman [81]	2018	Opinion	Improving career outcomes for technical education students	New technologies as a result of 4th Industrial Revolution
Barthauer [6]	2019	$N = 385$, surveys	Career sustainability of German scientists	Role misfit, burnout
Chin [16]	2019	$N = 2$, case studies	Career sustainability of Chinese manufacturing workers	New technologies, new business models
Heslin [35]	2019	Opinion	Role of learning mode in a sustainable career	Under-employment, lay-offs, unemployment
Jacobs [42]	2019	$N = 1874$, surveys	Role of networking for European freelancers	Organizations increasingly shift career development responsibilities to employees and other workers
Mutter [56]	2019	$N = 21$, interviews	Impacts on careers of stay-at-home couples	Sacrificing one's career at home while their partner is globally mobile (e.g. merchant seamen)
Peters [62]	2019	$N = 98$, surveys	Effects of age-based stereotyping on older supermarket workers	Negative stereotyping of older workers
Spooner [73]	2019	$N = 65$, interviews	Influence of training on career intentions of future General Practitioners	Rising workloads, isolation, ill-prepared to manage periphery activities to one's profession (e.g. financial management)
Stuer [75]	2019	$N = 5205$, surveys	Career sustainability of Belgian workers across the lifespan	Finding and retaining work
Xiao [87]	2019	$N = 614$, surveys	Effects of stock ownership on career development in China	Stress due to changes

Therefore, a sustainable career is more likely if one takes time for periodic renewal, one is flexible and adaptable to leverage new opportunities and one's career is holistically integrated to achieve an acceptable work-life balance trajectory. We find in general career research literature, career sustainability drivers help guide workers and human resource management professionals to build and maintain sustainable careers. But what about sustainable careers in project management?

Table 5.4 Sustainable career drivers

Author	Year	Empirical	Focus	Career sustainability drivers
Bennett [8]	2009	$N = 239$, surveys	Career sustainability for professional musicians and dancers	Develop small business skills (including marketing), teaching, adopt new technologies
Murray [55]	2009	Opinion	Improving the Australian political class	Improve career sustainability through continuous training and professionalism
Singhvi [71]	2010	Opinion	Building a sustainable career in science	Network, review papers, submit grants
Stronach [74]	2010	$N = 14$, mixed	Career sustainability of Indigenous Australian boxers	Pursue training and or education for renewal, work with mentors
Hall [31]	2012	$N = 73$, surveys	Career success while sustaining personal and family well-being	Understand that career success can be personal and subjective
Herman [34]	2012	$N = 38$	Motherhood and sustainable career in science, engineering, and technology	Develop a sense of entitlement for reduced working hours without forfeiting one's career
Watson [82]	2012	$N = 36$, interviews	Career sustainability of Victoria musicians	Develop a companion career related to the primary career, work with a mentor
Asuquo [4]	2013	$N = 210$, surveys	Employability of Nigerian youth	Learn continuously, network, be a team-player, be persistent, plan, take risks, be optimistic, be flexible
Siegel [70]	2013	Opinion	Career sustainability of orthopaedic surgeons	Manage yourself and your business, develop good relationships with vendors, maintain an active status with your medical licenses
Montgomery [54]	2014	Opinion	Mentoring graduate students and junior faculty	Work with a mentor over one's total career

(continued)

Table 5.4 (continued)

Author	Year	Empirical	Focus	Career sustainability drivers
Dahl [19]	2015	$N = 10$, interviews	Career sustainability of Swedish Leisure-time teachers	Develop profession-specific threshold competencies: developmental and group psychology, knowledge about social interaction, leadership and conflict management
Beeching [7]	2016	$N = 33$, interviews	Career sustainability for professional musicians	Know your audience, role flexibility, perform multiple roles
Hlanganipai [37]	2016	$N = 85$, surveys	Role of training and development in South African retail and wholesale managers	Plan training and development for your career
Rosenblum [68]	2016	Opinion	Professional identity and the clinician-scientist	Have certainty of personal values, ambitions, abilities; develop resilience, self-esteem and self-perceived competence; have tolerance for ambiguity and complexity; have the ability to manage complexity; work with a mentor; pursue life-career balance
De Vos [20]	2017	Opinion	Career sustainability in general	Plan personal career development rather than vertical advancement; manage work and non-work activities to minimize spillover; develop career competencies (self-awareness, self-management, etc.)
Van Driel [80]	2017	$N = 18$, interviews	Retaining Australia General Practitioner research leaders	Develop mentorships, alliances and, collaborations; take leadership training

(continued)

Table 5.4 (continued)

Author	Year	Empirical	Focus	Career sustainability drivers
De Vos [21]	In press	Opinion	Career sustainability model	Be proactive
Ko [45]	2018	Opinion	Career development for general surgeons	Be a competent generalist, but also specialize (dual careers)
Pajic [60]	2018	$N = 314$, surveys	Hungarian nurses career adaptability	Have a proactive and conscientious personality; be adaptable
Taponen [77]	2018	$N = 2613$, surveys	Career employment status and changes to due to asthma	Get advice from a counsellor; manage your asthma treatment
Barthauer [6]	2019	$N = 385$, surveys	Career sustainability of German scientists	Develop a long-term focus and consider career mobility
Chin [16]	2019	$N = 2$, case studies	Career sustainability of Chinese manufacturing workers	Become resourceful, flexible and integrative, plan for renewal
Frie [26]	2019	$N = 10$, interviews	How "flexperts" renew their expertise	Gain the ability to make sense of new knowledge and skills, and have confidence for mastery
Goodwin [29]	2019	$N = 28$	Developing self-efficacy and career optimism in Australian creative industries	Join a community of practice
Heslin [35]	2019	Opinion	Role of learning mode in a sustainable career	Be in the learning mode (always learning); apply self-directed and self-regulated learning
Hirschi [36]	2019	$N = 2679$, surveys	Sustainable careers and the impact of nonwork	Develop professional traits: agreeableness, extraversion and openness; and motives: curiosity and an interest in new experiences
Jacobs [42]	2019	$N = 1874$, surveys	Role of networking for European freelancers	Network proactively; take ownership for one's career

(continued)

Table 5.4 (continued)

Author	Year	Empirical	Focus	Career sustainability drivers
Kelly [43]	2019	$N = 280$, surveys	Role of leisure seriousness and sustainable careers	Engage in serious leisure; network while participating in leisure activities
Kotecha [48]	2019	Opinion	General practitioner career sustainability advice	Learn continuously; join a community of practice; lead professional conversations; manage your well-being; shape the future (innovation and research)
Merritt [53]	2019	Opinion	Building capacity of early career researchers	Develop threshold competencies (e.g. presentation skills, teamwork, etc.)
Mutter [56]	2019	$N = 21$, interviews	Impacts on careers of stay-at-home couples	Negotiate and plan for career development; look for flexible work agreements
Olsson [57]	2019	$N = 15$, interviews	Specialty choice of Swedish medical doctors	Be invited into your community of practice; fit in and contribute to your organization
Pajic	2018	$N = 314$, surveys	Hungarian nurses career adaptability	Have a proactive and conscientious personality; be adaptable
Suhairom [76]	2019	$N = 9$, interviews	Sustainable gastronomic careers	Develop career-related technical, business, and soft competencies
Bozionelos [9]	In press	$N = 334$, surveys	Career sustainability of Hong Kong salespeople	Become a continuous learner, open and adaptable; manage change
Chudzikowski [17]	In press	$N = 34$, interviews	Career sustainability of international management consultants	Conform and align early with one's organization; take ownership for own's career

5.3.3 Sustainable Project-Oriented Careers

We systematically review the six, project management-related journals and search everywhere for the term “sustainable career”, and we find only one result (see Table 5.5). We also search for the term “career” everywhere, and then again, but only in the title; and find that career sustainability is a new term in the project management-oriented literature. We review these papers for their relevance to sustainable careers. We learnt that many of the results were due to the authors using the term “career” publication biography, rather than somehow relating to their research.

Again, we find that within the project management literature, there were hints of sustainable career turbulence (see Table 5.6) and drivers (see Table 5.7).

When one reviews the project management literature, research and opinion articles about sustainable careers in project management do not exist; however, there is useful information related to sustainable project-oriented career turbulence and drivers. We believe that a perceptive project participant may see signs that their project-oriented career is unsustainable: that is, there may be “error messages” that one’s career is in trouble.

5.4 Error Messages Your Career Is in Trouble

We offer error messages as indicators that one’s career may not be sustainable. Hartman [32] developed a series of project management error messages designed to alert the project manager that something is wrong in their project. While no one sustainable project-oriented career error message may be the “nail in the coffin”, a sustainable project-oriented career is in jeopardy as error messages accumulate. After our literature review analysis, we have categorized these error messages into (i) renewal opportunities, (ii) adaptability and (iii) holistic integration (e.g. work-life balance).

Table 5.5 Sustainable careers in project management literature

Journal	Sustainable career everywhere	Career everywhere	Career in title
International Project Management Journal	0	397	25
Project Management Journal	1	279	3
Project Management Research and Practice	0	4	0
Modern Project Management	0	1	0
International Journal of Construction Management	0	11	0
Journal of Engineering, Project and Production Management	0	10	0

Table 5.6 Turbulence in project-oriented sustainable careers

Author	Year	Empirical	Project mgmt focus	Turbulence
Bredin [10]	2013	$N = 10$, case study	Project managers and career models of Swedish project-oriented firms	Lack of job rotation to increase competency
Crawford [18]	2013	$N = 9$, interviews	Project career paths in Australia	Rise of contract work
Havermans [33]	2019	$N = 21$, interviews	Explore the motivation and experience of becoming a project manager	Frustrating processes, project administration, stress from conflict and work overload, changing work conditions, accountability without authority, politics
Huemann [39]	2019	$N = 20$, Interviews	Project manager career development and line manager relationships	Lack of project manager, career development support from line managers

5.4.1 *Error Messages: Lack of Flexibility*

Assigned the same types of projects time after time,
Haphazard and infrequent networking to uncover new opportunities,
Did not get the promotion, job or contract,
Shut out from normal activities (e.g. not invited or contracting meetings),
Works on the same aspects of PM on all projects (scheduling, e.g. without cross-fertilization of competence in other knowledge areas),
Does not contribute to professional activities outside the workplace,
Does not promote the company he or she is working for in meetings or conferences outside of the organization,
Does not update the work procedures according to the lessons learned or from reading new PM literature,
Can only work with particular types of people,
Not open to new ideas coming from subordinates or the project team.

These types of error messages may indicate that your career may benefit from being open to new opportunities to expand career flexibility.

5.4.2 *Error Messages: Lack of Holistic Integration*

Clinically depressed,

Table 5.7 Project-oriented sustainable career drivers

Author	Year	Empirical	Project mgmt focus	Career sustainability drivers
Ahsan [2]	2013	$N = 795$, content analysis	Comparative analysis of competencies in job advertisements	Have the necessary competencies called for in the job advertisement
Bredin [10]	2013	$N = 10$, case study	Project managers and career models of Swedish project-oriented firms	Work for organizations that have career models to guide project managers' career development; project management training programs; offer organizational career development opportunities; and have a talent management program to build project management capacity; build personal project management competencies
Crawford [18]	2013	$N = 9$, Interviews	Project career paths in Australia	Show movement from simpler to more complex projects; work within collaborative and innovative project structures; engage with mentors
McKevitt [52]	2017	$N = 207$, surveys	Work identify and career satisfaction of IT project managers	Develop a high project management identity that can lead to career satisfaction
Gruden [30]	2018	$N = 69$, surveys	Influence of behavioural competencies on project performance	Develop behavioural competencies
Chen [15]	2019	$N = 228$, surveys $N = 10$, interviews	Identify relevant project management competencies for Chinese construction companies	Follow a career path of increasing responsibility
Havermans [33]	2019	$N = 21$, interviews	Explore the motivation and experience of becoming a project manager	Have the necessary competencies to successfully manage the project; be aware that there may be conflicting expectations of superiors
Huemann [39]	2019	$N = 20$, Interviews	Project manager career development and line managers	Work with a mentor; have a buddy; and, participate in a community of practice
Wen [83]	2019	$N = 163$ m survey	Project manager's competencies in the project closing phase	Get feedback at closeout as an input to career planning

No longer exercise as often as you would like,
Unplanned career absences,
High annual leave balance,
Does not take advantage of employer-offered stress avoidance or stress management schemes or employee assistance programmes,
Regularly works more than 50 h per week,
Regularly works at home after a full day of work for more than an hour or more than two evenings a week,
Consistently works at least half a day on weekends for more than one weekend a month,
Thinks about the undesirable things that occurred in the workplace,
Stressed about the risks that are extremely unlikely to happen.

These types of error messages may indicate that your career may benefit from more work-life balance to have a more holistic integration.

5.4.3 Error Messages: Lack of Renewal Opportunities

Long time since your last formal class that had an assessment component,
Infrequent professional development other than mandatory learning requirements (e.g. Fire Safety),
Unfamiliar with Adaptive Project Management methods (Agile, Scrum, #NoProjects, Kanban, Lean),
Curriculum vitae is out of date,
Lack of professional institution membership (e.g. IPMA, PMI, etc.),
Infrequently reads journals or magazines from your industry,
Headhunters seldom contact you,
Infrequent and irregular online professional social network participation (i.e. LinkedIn, ProjectManagement.com, GanttHead.com, etc.),
Unfamiliar with emerging PM computer/mobile applications. (i.e. Slack, Monday, WorkflowMax, Wrike, Trello, Mavenlink, Asana, etc.).

These types of error messages may indicate that a sustainable project-oriented career may benefit from an intervention such as renewal. At the same time that we see disruption and transformation in project management, we also see a great change in education and training that will be of interest to those who seek formal renewal.

5.5 Renewal: Innovations in Training and Education

One can renew their skill set through education and training. Training involves attaining specific and practical skills, usually through relatively short-term learning experiences. Education involves more than theoretical learning in a classroom setting about a wide variety of topics gained over a relatively long term. Training often prepares a learner for the present while education prepares an individual for a future job [67]. While the focus of this paper is about preparing for the future by having a sustainable career, we do want to emphasize the importance of training in project management. Project-oriented participants will do well to review the wide variety of training opportunities offered by professional organizations like the International Project Management Association, Project Management Institute, Association for the Advancement of Cost Engineering International, British Computer Society and others.

There has been a significant change to the higher education sector that better support successful renewal opportunities:

1. *Multiple Degree Options*: There is greater diversity within degrees so that a learner can find a better fit with the degree and their goals [47]. Take the Master of Business Administration (MBA) degree, it was not long ago that an MBA was generic rather than specialized. Now, one can get an MBA in Project Management, MBA in Project and Operations Management, MBA in Business Project Management, etc.
2. *Combined Degrees*: Universities increasingly offering combined degrees (e.g. a Master of Project Innovation and Master of Project Management) where the learner graduates with two degrees in a shorter period since electives and core classes are shared between the two degrees.
3. *Accelerated Programmes*: Universities are finding ways to deliver the full degree but within compressed time frames so that the learner achieves the learning outcomes but over a shorter period [44]. For example, some universities offer classes over three rather than the typical two semesters per year, resulting in students graduating sooner.
4. *Intensive Classes:* Students can attend intensive classes (e.g. Thursday, Friday and Saturday) so that they can work most of the week and attend classes at the end of the week and on the weekend. Intensive classes improve access to educational renewal [47].
5. *Problem-Oriented*: More educational programmes are supplementing theoretical learning with problem-oriented thinking supplemented with design thinking [49]; that is, they use newly acquired theory to solve problems. By doing so, learning is reinforced.
6. *Personalized Learning*: Students have more choice within courses to personalize learning and more electives to achieve their learning outcomes often through academic mentorship [14].
7. *Interactive*: More and more, lectures are replaced with interactive learning [47] activities like discussions, role plays, peer to peer, board rotation, etc.
8. *Flipped Classroom*: The flipped classroom is a relatively new teaching approach where the traditional classroom activities are flipped with students complete assigned learning activities before they attend the class. The key advantage is

that students in a flipped classroom, have more guidance with higher order learning activities like analysis and evaluation, to improve learning.

9. *Outdoor Learning Spaces*: Universities are adding outdoor learning spaces to bring additional diversity to the educational environment [13]. Rather than self-directed group learning and discussions occurring in classrooms, this type of learning environment is especially welcomed by adults who enjoy getting out of the classroom to learn.
10. *Workplace Readiness/Authentic Assessment*: Perhaps one of the largest changes is that many university programmes have adapted their curriculum so that the content is more likely to be used in practice [14, 47]. Students increasingly ask that they become workplace-ready as they progress through their program. This feedback has triggered a rethinking of how to assess student learning. This might mean a decline in exams and essays, and an increase in reports, problem-solving exercises and other types of assignments to approximate what is done in the real world. More institutions are addressing workplace readiness and including authentic assessment [84].
11. *Digital Badges*: Digital badges (micro-credentials) are an emerging online system to recognize accomplishments and competence [22]. An awarding organization offers a learning opportunity for a student. Successful students are awarded a digital badge to recognize their competence. The online badge includes key information such as badge name and description, awarding criteria, issuer, learning evidence, date issued, etc. Some universities issue digital badges (e.g. Design Thinker) after one or two subjects are completed to a certain standard (e.g. 65%). The student can add this digital badge to their CV and to an online platform like LinkedIn rather than waiting until graduation to show the fruit of their efforts.

Therefore, while there is great change occurring in the project environment, there are also increased opportunities and modes of learning for those project-oriented participants who are looking for renewal opportunities to boost their sustainable project-oriented career.

5.6 Conclusion

Long gone is the norm that one had a job for a lifetime—a stable, singular and linear career. More project-oriented participants understand the simple concept of supply and demand: if one has competencies that are in demand, then one will likely remain employed. Organizations are becoming more project-oriented and recruit project participants often for particular jobs, and may retain them for additional project work or release them [38]. Many work arrangements are fragile, given changes brought about to how we work (disruptive technologies brought about by the 4th Industrial Revolution) and how we organize for work (more outsourcing and contract work in the Gig Economy). Add work changes to career turbulence external shocks (e.g.

family commitments or job promotion), then a project-oriented participant may feel out of equilibrium with an unsustainable project-oriented career. One may have a respected project-oriented job but may need to work long hours to meet deadlines, perhaps resulting in family conflict, stress and possible burnout. Such imbalance is usually not sustainable. One may also have a good project-oriented job but has not expanded or renewed their skill set to be able to contribute to projects involving disruptive technologies. Or they may have a singular career focus, rather than a composite career, with multiple, and perhaps entrepreneurial initiatives.

Career sustainability is complex, non-linear and fluid; career sustainability is more likely if one addresses at least three key elements: (i) holistic integration with one's life, (ii) career flexibility and (iii) career renewal [21]. Compounding career planning and development is that one's personal goals change over time [21]; the sports car that was important in one's twenties becomes less important with marriage and the arrival of twin babies! During one's career, there may be changes in the degree of career fit and sustainability [61]. Hirschi et al. [36] advise to take a whole-life perspective when living a sustainable career; we might be guided by sustainable career drivers and unsustainable career error messages in order to take proactive planning and action. Unsustainable career error messages can inform us if any of these areas are weak. Career sustainability is of even greater importance to some given that we have an ageing population that is expected to work longer in their careers [75].

We conclude this paper with an overview of new trends occurring in education that make renewal more attractive and feasible for most. Project-oriented career sustainability is the individual's responsibility according to a protean view; there are positive actions one can take to have a sustainable career in projects.

The general area of career sustainability research is in its infancy [51]; some call for more research about the nature and dimensions of sustainable careers, indicators of a sustainable career and general sustainable career theory building [21]. However, it appears to this research team that the research opportunities in project-oriented careers are broad and deep.

References

1. Adams RM (2006) Sustainable career management in the pharmaceutical industry—lessons and insights. JMMDDPM 6(1):74–76
2. Ahsan K, Ho M, Khan S (2013) Recruiting project managers: a comparative analysis of competencies and recruitment signals from job advertisements. PMJ 44(5):36–54
3. Akkermans J, Seibert SE, Mol ST (2018) Tales of the unexpected: integrating career shocks in the contemporary careers literature. SAJIP 44(6):1–10
4. Asuquo PN, Inaja AE (2013) Fostering sustainable career development and employability among young people in the changing world of work: employers' perspective. Procedia: SBS 84:1492–1499
5. Avis J (2018) Socio-technical imaginary of the fourth industrial revolution and its implications for vocational education and training: a literature review. JVE&T 70(3):337–363
6. Barthauer L, Kaucher P, Spurk D, Kauffeld S. Burnout and career (un)sustainability: looking into the blackbox of burnout triggered career turnover intentions. J Voc Behav (in press)

7. Beeching AM (2016) Who is audience? Arts Humanit High Educ 15(3–4):395–400
8. Bennett D (2009) Academy and the real world: developing realistic notions of career in the performing arts. Arts Humanit High Educ 8(3):309–327
9. Bozionelos N, Lin C, Lee KY. Enhancing the sustainability of employees' careers through training: The roles of career actors' openness and of supervisor support. J Voc Behav (in press)
10. Bredin K, Söderlund J (2013) Project managers and career models: an exploratory comparative study. IJPM 31(6):889–902
11. Calderón A, Ruiz M (2015) A systematic literature review on serious games evaluation: an application to software project management. Comput Educ 87:396–422
12. Carayannis EG, Kwak YH, Anbari FT (2005) The story of managing projects: an interdisciplinary approach. ABC-CLIO Santa Barbara, California
13. Cassidy A, Wright A, Strean WB, Watson G (2015) The interplay of space, place and identity: transforming our learning experiences in an outdoor setting. Collected Essays Learn Teach 8:27–34
14. Celuch K, Bourdeau B, Khayum M, Townsend L (2017) The role of the university in accelerated learning and innovation as a regional ecosystem integrator. JRIT&L 10(1):34–47
15. Chen T, Fu M, Liu R, Xu X, Zhou S, Liu B (2019) How do project management competencies change within the project management career model in large Chinese construction companies? IJPM 37(3):485–500
16. Chin T, Li G, Jiao H, Addo F, Jawahar IM (2019) Career sustainability during manufacturing innovation. Career Dev Int 24(6):509–528
17. Chudzikowski K, Gustafsson S, Tams S. Constructing alignment for sustainable careers: insights from the career narratives of management consultants. J Voc Behav 110:1 (in press)
18. Crawford L, French E, Lloyd-Walker B (2013) From outpost to outback: project career paths in Australia. IJPM 31(8):1175–1187
19. Dahl M, Karlsudd P (2015) Leisure-time teachers in a changed profession. Probl Educ 21st Century 68:22–35
20. De Vos A, Van Der Heijden, Beatrice, Ijm (2017) Current thinking on contemporary careers: the key roles of sustainable HRM and sustainability of careers. Curr Opin Environ Sustain 28(41–50)
21. De Vos A, Van Der Heijden B, Akkermans J. Sustainable careers: towards a conceptual model. J Voc Behav (in press)
22. DiSalvio P (2016) New pathways to credentialing: the digital badge. https://nebhe.org/journal/new-pathways-to-credentialing-the-digital-badge/ Accessed 30 Jan 2020
23. Erdogmus T, Czermak M, Baumsteiger D, Kohn D, Boller-Hoffecker A, Schmidt N, Linden R (2018) How to support clients and vendors in IT outsourcing engagements: the different roles of third-party advisory services. JITTC (8)2:184–191
24. Favaro K (2015) A brief history of the ways companies compete. HBR 11–21
25. Fournier, Geneviève, Zimmermann, Hélène, Masdonati Jonas, Gauthier Christine (2018) Job loss in a group of older Canadian workers: challenges in the sustainable labour market reintegration process. Sustainability 10(7):2245
26. Frie Lonneke S, Potting, Karin CJM, Sjoer Ellen, Van Der Heijden, Beatrice IJM, Korzilius, Hubert PLM (2019) How flexperts deal with changing expertise demands: a qualitative study into the processes of expertise renewal. HRDQ 30(1):61–79
27. Gartner (2019) Predicts 2019: establish the foundations for next-generation digital government success. Gartner INC
28. Gemunden HG, Lehner P, Kock A (2018) The project-oriented organization and its contribution to innovation. IJPM 36:147–160
29. Goodwin K (2019) Developing self-efficacy and career optimism through participation in communities of practice within Australian creative industries. AJCD 28(2):122–131
30. Gruden N, Stare A (2018) The influence of behavioral competencies on project performance. PMJ 49(3):98–109
31. Hall DT, Lee MD, Kossek EE, Heras ML (2012) Pursuing career success while sustaining personal and family well-being: a study of reduced-load professionals over time. JSI 68(4):742–766

32. Hartman FT (1999) Don't park your brain outside: a practical guide to improving shareholder value with SMART management. Project Management Institute, Newtown Square
33. Havermans Liselore, Van Der Heijden, Beatrice IJM, Savelsbergh Chantal, Storm Peter (2019) Rolling into the profession: exploring the motivation and experience of becoming a project manager. PMJ 50(3):346–360
34. Herman C, Lewis S (2012) Entitled to a sustainable career? Motherhood in science, engineering, and technology. JSI 68(4):767–789
35. Heslin PA, Keating LA, Ashford SJ. How being in learning mode may enable a sustainable career across the lifespan. J Voc Behav (in press)
36. Hirschi A, Steiner R, Burmeister A, Johnston CS. A whole-life perspective of sustainable careers: the nature and consequences of nonwork orientations. J Voc Behav (in press)
37. H langanipai N, Musara M (2016) Training and development for career management and talent development in wholesale and retail industry in South Africa. J Psychol Afr 26(2):164–166
38. Huemann M, Keegan A, Turner JR (2007) Human resource management in the project-oriented company: a review. IPMJ 25(3):315–323
39. Huemann M, Ringhofer C, Keegan A (2019) Who supports project careers? Leveraging the compensatory roles of line managers. PMJ 50(4):476–486
40. Ikediashi D, Aigbavboa C (2019) Outsourcing as a strategy for facilities management provision in Nigerian universities. Int J Const Manag 19(4):281–290
41. Iles P (1997) Sustainable high-potential career development: a resource-based view. Career Dev Int 2(7):347–353
42. Jacobs S, De Vos A, Stuer D, Van der Heijden Beatrice IJM (2019) Knowing me, knowing You' the importance of networking for freelancers' careers: examining the mediating role of need for relatedness fulfillment and employability-enhancing competencies. Front Psychol 10:1–14
43. Kelly CM, Strauss K, Arnold J, Stride C. The relationship between leisure activities and psychological resources that support a sustainable career: the role of leisure seriousness and work-leisure similarity. J Voc Behav (in press)
44. Kitchener IL (2017) Are accelerated degrees the future for higher education? Compass J Learn Teach 10(3):1–4
45. Ko B, Mchenry CR (2018) A model for a career in a specialty of general surgery: one surgeon's opinion. Am J Surg 215(1):8–13
46. Kobie N (2018) What is the gig economy and why is it so controversial? https://www.wired.co.uk/article/what-is-the-gig-economy-meaning-definition-why-is-it-called-gig-economy. Accessed 30 Jan 2020
47. Kofinas A, Bentley Y, Minett-Smith C, Cao G (2017) Block teaching as the basis for an innovative redesign of the PG suite of programmes in University of Bedfordshire Business School. Paper presented at third international conference on higher education advances, Editorial Universitat Politècnica de València, Valencia, Spain, 21–23 June 2017
48. Kotecha A (2019) Crammer's corner: life after certificate of completion of training. SAGE Publications, London, England
49. Linton G, Klinton M (2019) University entrepreneurship education: a design thinking approach to learning. JIAE 8(1):1–11
50. Mayrhofer W (2016) Going the distance: contextualizing and examining the sustainable career. AOM Proc 2016(1):14372
51. Mcdonald KS, Hite LM (2018) Conceptualizing and creating sustainable careers". HRDR 17(4):349–372
52. McKevitt D, Carbery R, Lyons A (2017) A profession but not a career? Work identity and career satisfaction in project management. IJPM 35(8):1673–1682
53. Merritt C, Jack H, Mangezi W, Chibanda D, Abas M (2019) Positioning for success: building capacity in academic competencies for early-career researchers in sub-Saharan Africa. GMH 6
54. Montgomery BL, Dodson JE, Johnson SM (2014) Guiding the way: mentoring graduate students and junior faculty for sustainable academic careers. SAGE Open 4(4):1–11

55. Murray A (2009) Can better political governance give Australia an improved political class? Agenda: JPAR 16(3):63–67
56. Mutter J, Thorn K (2019) Global mobility and the career of the stay-at-home partner. JGM 7(1):88–102
57. Olsson C, Kalén S, Mellstrand Navarro C, Ponzer S (2019) Swedish doctors' experiences and personality regarding medical specialty choice: a qualitative study. Int J Med Ed 10:36
58. Osif BA (2009) Work/life balance. LL&M 23(1):42–46
59. Overbaugh J (2011) 24/7 isn't the only way: a healthy work–life balance can enhance research. Nature 477(7362):27
60. Pajic S, Keszler Á, Kismihók G, Mol ST, Den Hartog D (2018) Antecedents and outcomes of Hungarian nurses' career adaptability. Int J Manpower 39(8):1096–1114
61. Parasuraman S, Greenhaus J, Linnehan F (2000) Time, person-career fit, and the boundaryless career. TOB 7:63–78
62. Peters P, Van Der Heijden B, Spurk D, De Vos A, Klaassen R (2019) Please don't look at me that way. An empirical study into the effects of age-based (meta-)stereotyping on employability enhancement among older supermarket workers. Front Psychol 10(249)
63. PMI (2019) The future of work: leading the way with PMTQ. Project Management Institute, Newtown Square
64. PMI (2018a) Maximizing the benefits of disruptive technologies on projects. Project Management Institute, Newtown Square
65. PMI (2018b) The project manager of the future: developing digital-age project management skills to thrive in disruptive times. Project Management Institute, Newtown Square
66. PMI (2017) Job growth and talent gap: 2017–2027. Project Management Institute, Newtown Square
67. Rickman P (2004) Education vs training. Phil Now 47. https://philosophynow.org/issues/47/Education_versus_Training. Accessed 30 Jan 2020
68. Rosenblum N, Kluijtmans M, Ten Cate O (2016) Professional identity formation and the clinician-scientist: a paradigm for a clinical career combining two distinct disciplines. Acad Med 91(12):1612–1617
69. Schwab K (2016) The fourth industrial revolution: what it means, how to respond. World Economic Forum. https://www.weforum.org/agenda/2016/01/the-fourth-industrial-revolution-what-it-means-and-how-to-respond/. Accessed 30 Jan 2020
70. Siegel J (2013) Getting organized in your job. J Orthop Trauma 27(1):8
71. Singhvi A, Sachdev P (2010) Building a sustainable career in science. Nat Biotechnol 28(4):378
72. Skulmoski G (2001) Project maturity and competence interface. Cost Eng 43(6):11–18
73. Spooner S, Laverty L, Checkland K (2019) The influence of training experiences on career intentions of the future GP workforce: a qualitative study of new GPs in England. BJGP 69(685):578
74. Stronach M, Adair D (2010) Lords of the square ring: future capital and career transition issues for elite indigenous Australian boxers. CCSJ 2(2):46–70
75. Stuer D, De Vos A, Van Der Heijden B, Akkermans J (2019) A career perspective of work ability: the importance of resources across the lifespan. IJERPH 16(14)
76. Suhairom N, Musta'amal AH, Mohammed Amin NF, Kamin Y, Abdul Wahid NH (2019) Quality culinary workforce competencies for sustainable career development among culinary professionals. Int J Hosp Manag 81:205–220
77. Taponen S, Lehtimäki L, Karvala K, Luukkonen R, Uitti J (2018) Employment status and changes in working career in relation to asthma: a cross-sectional survey. JOMT 3(1):8
78. Valcour M (2013) Craft a sustainable career. HBR 15 Jul 2013
79. Van Der Heijden B, De Vos A, Akkermans J, Spurk D, Semeijn J, Van Der Velde M, Fugate M. Sustainable careers across the lifespan: moving the field forward. J Voc Behav (in press)
80. Van Driel M, Deckx L, Cooke G, Pirotta M, Gill GF, Winzenberg T (2017) Growing and retaining general practice research leaders in Australia: how can we do better? Aus Fam Physician 46(10):757

81. Venkatraman S, de Souza-Daw T, Kaspi S (2018) Improving employment outcomes of career and technical education students. HESWBL 8(4):469–483
82. Watson A, Forrest D (2012) The bands culture in Victoria, Australia: live music benefits career paths, employment and community. AJME 2:71–81
83. Wen Q, Qiang M (2019) Project managers' competences in managing project closing. PMJ 50(3):361–375
84. Wiewiora A, Kowalkiewicz A (2019) The role of authentic assessment in developing authentic leadership identity and competencies. Assess Eval High Ed 44(3):415–430
85. Wilkinson M, Harper B, Peacock J, Morrison R, Getz K (2019) Assessing outsourcing oversight practices and performance. TIRS 1:1–9
86. Witzig TE, Smith SM (2019) Work-life balance solutions for physicians-It's all about you, your work, and others. Mayo Clin Proc 94(4):573–576
87. Xiao H, Shi Y, Varma A (2019) The effects of employee stock ownership plans on career development in a new era. Career Dev Int 24(5):453–474

Chapter 6
Environmental Project Management: Contextual Enablers and Boundaries

Vladimir Obradović, Marija Todorović, and Milica Jovanović

Abstract This paper aims to identify the factors that are contributing to or limiting project management application in the field of environment protection at the national and local levels. Based on those factors, the paper aims to define key enablers and important actions to integrate project management in environmental projects. The project management theory in the context of environmental protection suggests that the strategy for implementing project management should be adjusted to local needs in terms of the economic, political, social and administrative system. Based on the literature review, empirical research has been conducted, supported with interviews with environmental officials from local government units in the Republic of Serbia. The primary conclusion finds collaboration with relevant local and national institutions, support from local government and the presence of the relevant strategic documents at the local level as the main influencing factors. The paper tends to define the current state of project management in this field and defines important actions to support the capacity building of local governments in the area of environmental project management.

Keywords Environment · Protection · Project · Local · Government

6.1 Introduction

In the last four decades, the global focus has changed from environmental politics to the politics of sustainable development. From the perspective of developing countries, the period after the 1992 conference in Rio de Janeiro is recognized as an era of

V. Obradović (✉) · M. Todorović · M. Jovanović
Faculty of Organizational Sciences, University of Belgrade, Jove Ilića 154, 11000 Belgrade, Serbia
e-mail: vladimir.obradovic@fon.bg.ac.rs

M. Todorović
e-mail: marija.todorovic@fon.bg.ac.rs

M. Jovanović
e-mail: milica.jovanovic@fon.bg.ac.rs

R. Cuevas et al. (eds.), *Research on Project, Programme and Portfolio Management*, Lecture Notes in Management and Industrial Engineering,
https://doi.org/10.1007/978-3-030-60139-3_6

growing involvement in the global environment protection build on the principles of sustainable development [2, 4].

Concerning environmental governance at the global, national and local level in the last ten years, a new opportunity to enhance economic growth through environmental projects has been seen [7].

With the innovation and technology boom at the global and local markets, the relation between environmental regulation and technological innovation has been proven. The new ecological legislation can become a unique opportunity for companies that would not otherwise perform these activities [9, 15, 22].

Beside direct effect on nature, environmental protection projects have an impact on the whole community. Explicit and tacit knowledge and experience gained from these projects can serve as a database for a broader spectrum of initiatives at the local and national levels. Employees in these predominantly donor-funded projects can gain knowledge about the procedures, planning, reporting and project leading. They also can gain practical technical expertise that can be used in land management, agriculture, etc. [3].

Besides direct impact on the ecological system, overall goals of each environment protection project influence social aspects, economy, attitudes and behaviours of the local community and its members [5, 19]. Adequate project management can influence future project and policy goals, and thus the well-being of many stakeholders [1, 14, 20, 21].

The primary purpose of this paper is to identify the factors that are contributing to or limiting project management application in the field of environment protection at the national and local levels. Based on those factors, the paper aims to define critical enablers and essential actions to integrate project management in environmental projects.

Environment protection projects are most often public projects, mainly carried out by public institutions. These institutions have none or only a small number of competitors as service providers, creating in that way a dominant position for a project owner and a project manager. The project environment is very complex: political influence, political changes, many stakeholders involved (agencies, local service providers, NGOs, companies), national institutions, donors, legislative framework [12, 18, 23]. Projects need to be sensitive to both the politics of scale and the politics of networks. These constraints emphasize project complexity, the importance of relationship management and user consultation [6, 8].

Environmental projects are mostly dependent on other projects or are a part of a more significant programme. To manage the programme at the local or national level, it is crucial to ensure funds as well as the commitment of the local authorities to the programme objectives and contracted agreement with donors. The project or programme success is heavily influenced by the project management maturity of an institution [11].

Furthermore, strategic orientation, vision and mission of the institution can make a high impact on people's lives, through the projects and programmes [10]. The project management theory in the context of environmental protection suggests that the strategy for implementing project management should be adjusted to local needs in terms of the economic, political, social and administrative system [13, 17].

According to Rwelamila [16], for a programme to fulfil its mandate successfully, addressing the organizational structure, personnel qualifications and programme management system are crucial elements. The study provided by Yanwen [24] confirms that the projects and programme management is limited and should be consistent with the cultural characteristics of a country (national and local institutions).

Main challenges of proper project management for environment protection are in the field of project processes: project initiation, planning, monitoring, reporting [21].

6.2 Methodology

6.2.1 Research Questions

Based on defined contextual boundaries and enablers identified based on systematic literature review, key research questions were focused on the following:

1. What are project management elements that are being applied in environmental projects?
2. Who are the main cooperation partners on environment protection projects?
3. What are the factors that are contributing to or limiting project management application in the field of environment protection?
4. What are the significant enablers and necessary actions to integrate project management in environmental projects?

6.2.2 Research Method

To investigate the current state of project management in this field and define necessary actions to support the capacity building of local governments (LG) in the area of environmental project management, a questionnaire was developed and distributed to the local governments in Serbia.

The first part of the questionnaire was composed of questions related to the demographic characteristics of respondents. The second part was dedicated to the standard elements of project management, where respondents were given the scale from 1 to 5 to assess the applicability of those elements on their projects.

The third part aimed to investigate the level of presence of collaboration with different actors at the local and national levels. The question was based on research provided by Hall and Holt [8], Bulkeley [6] and Santos and Varajão [18].

In addition to the above mentioned, respondents expressed their opinion on the recognized issues related to project management in the environmental protection arena. They were given a list of possible challenges and boundaries that was based on a primary conclusion from the research in this field [6, 8, 10, 11, 18, 21].

The fourth part of the survey focuses on the developing of the institutional capacities to support environmental protection project management.

This empirical research was supported with interviews with LSGs officials for environment protection, aiming to define a more profound interpretation of the responses. The results will be presented and discussed, along with the survey results in the section below.

6.3 Results and Discussion

6.3.1 Sample Description

The research was conducted during the first half of 2019 among local environmental officers from more than 40 municipalities in Serbia. This research was undertaken within the project "Capacity Building on Environmental Project Design, Proposal Development and Project Management for Local Government Units"

The preliminary analysis focused on the logical check of the response, and hence, incomplete questionnaires were not considered for further study. Therefore, only 63 questionnaires have been taken considered, for which 64.5% of the respondents were a core project team member, 16.5% were project managers and 19% were project associates.

This research has started questioning the essential characteristics of projects for environment protection at the local level. The respondents were mainly leading or participating in infrastructure projects, waste management project, capacity building projects and education (Fig. 6.1).

Through interviews with LGs' officials for environmental protection, it was elaborated that LGs in most of the cases need to solve infrastructure issues first. The reason

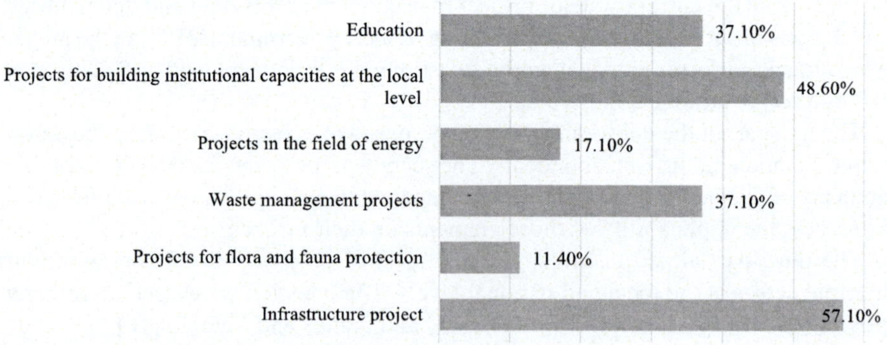

Fig. 6.1 Types of environment protection projects

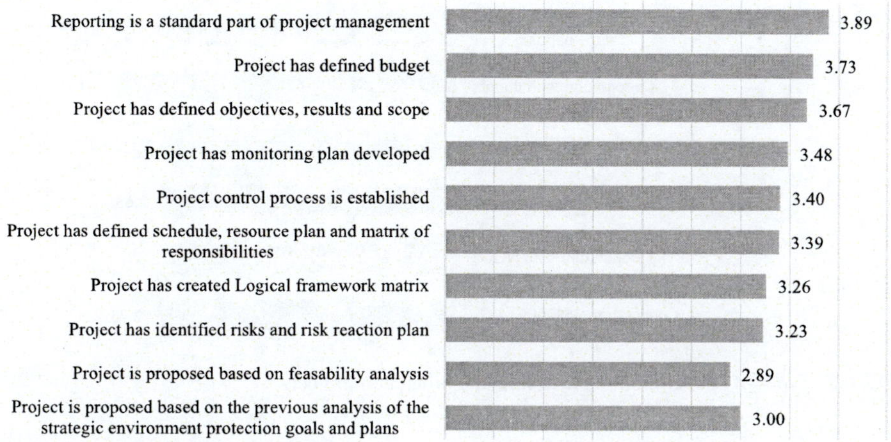

Fig. 6.2 Project management elements

for the excessive number of projects for waste management is the new national Law on Waste Management, and the various educational projects have a substantial impact in comparison to the funds allocated.

Before one could define what the key enablers and boundaries to project management in the field of environmental protection are, authors tended to investigate the current state of project management in LGs. On a scale from 1 to 5, the respondents ranked the level of project management elements on their projects. Results showed that the project initiation phase, as the most critical one (the alignment with strategic documents, goals and plans at the local and national level) is not the part of a practice. Logical framework matrix as a tool that is linking overall goals, specific objectives with project results is also rarely used.

Project plan in terms of schedule, budget and resources is mostly present, as well as reporting due to the fact that all expenditures must be justified to the local and national government or donors (Fig. 6.2). Interviews showed that this is not a planned activity but an obligatory one.

The project often requires cooperation with institutions at the local and national level since environmental protection projects have many stakeholders who make these projects very complex. Beside institutions at the local level, some decisions depend on the responsible ministry and other national institutions. Project success is also affected by cooperation with the NGO sector and companies at the local level, relationships with donors and relation with other municipalities to gain know-how or to jointly apply for a project.

Research results showed that communication and cooperation with local and national institutions is the most highly ranked type of collaboration (Fig. 6.3). The cooperation with NGOs is at a very low level. Interviews showed that LGs officials are not considering NGOs as a potential partner in service providing.

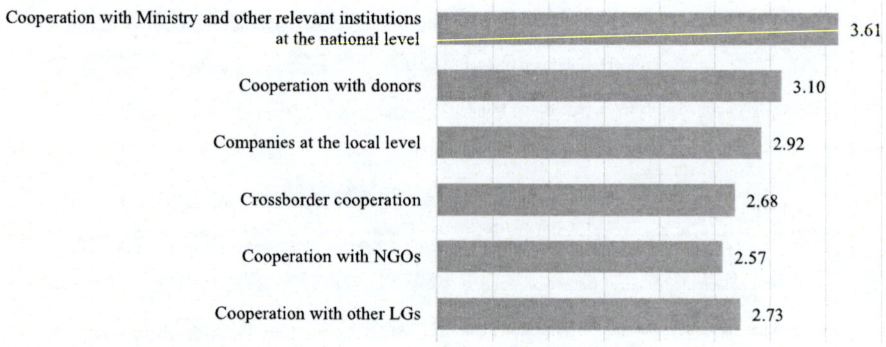

Fig. 6.3 Types of cooperation in environmental protection projects

One of the research questions was oriented towards factors that are contributing to or limiting project management application in the field of environment protection. The respondents had the opportunity to assess the presence of listed challenges and boundaries on a scale from 1 to 5. As explained in the section above, the list of specific challenges was created based on a literature review in the field of environment protection. According to the results, the most challenging is the attitude of people and the setting, which very often are not familiar with result-oriented work (Fig. 6.4).

Through desk analysis and in-depth interviews, one may conclude that environmental protection projects are in high per cent public projects, owned by the municipality or other public institution. Further, the governmental sector and institutions in Serbia are complex and in a phase of an on-going reform, where a process-oriented work is a dominant way of practice. Respondents did not recognize a lack of

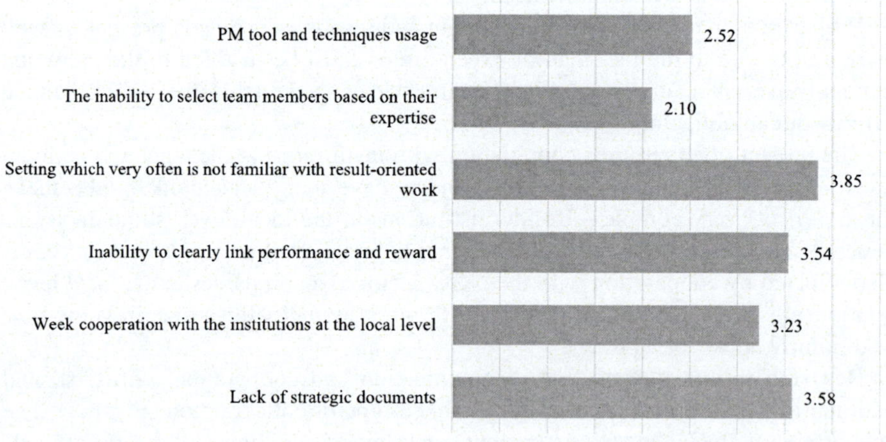

Fig. 6.4 Factors that are contributing to or limiting project management application in the field of environment protection

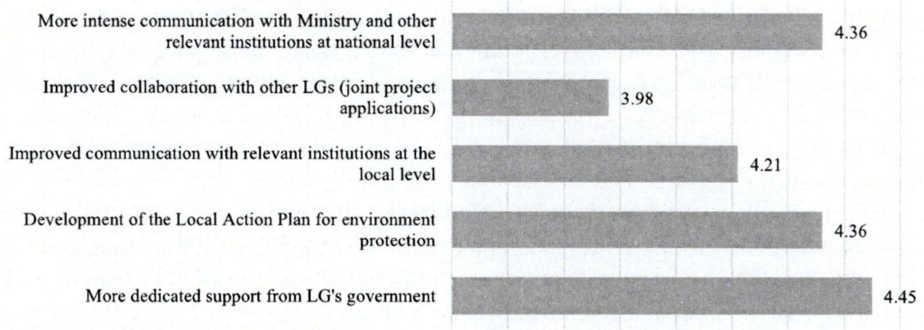

Fig. 6.5 Key enablers and important actions to integrate project management in environmental projects

project management tools and techniques application as challenging, which means that people and perspective aspects are the critical boundaries to project management.

Based on factors that are limiting project management application in this field, the respondents were asked to assess the importance of the enablers and actions to integrate project management in the environmental projects.

From the results, one can conclude that projects need more dedicated support of LGs government. Interviews showed that this is related to political influence, changes in political orientation, and key decision-makers.

Moreover, to improve project management in this field, LGs need to develop strategic documents in the environmental protection area and improved communication with the institutions at the national level (Fig. 6.5). The interview showed that LGs need strategic documents (sectors strategies and local action plans) to initiate, argument and prioritize projects.

6.4 Conclusion

The primary purpose of this paper was to identify the factors that are contributing to or limiting project management application in the field of environment protection at the national and local levels and based on these factors to define vital enablers and critical actions to integrate project management in environmental projects. The paper presents the results of the research conducted in local governments of the Republic of Serbia, as key institutions accountable for these projects in the local communities.

The paper defines and describes the current state of project management in this field. Results showed that the alignment with strategic documents, goals and plans at the local and national levels, in most of the cases are not part of standard everyday practice. Project plan in terms of schedule, budget and resources is mostly present, as well as reporting because budgetary expenditures must be justified in line with local and national legislative and donor procedures.

The key findings are that collaboration with relevant local and national institutions, dedicated support from LGs' government and the presence of the adequate strategic documents at the local level are the main enablers for project management application in environment protection.

These steps should build capacities of LGs in the area of environmental project management.

This research is relevant because it addresses the subject of vital contextual enablers and boundaries for project management in the field of environment protection which lately has a significant impact on social and economic growth and sustainable development in the much broader context.

Acknowledgements This paper is a result of the research project entitled "Research of Institutional and Personal Capacities in The Field of Project Management in the Area of Environmental Protection" implemented by Serbian Project Management Association—IPMA Serbia.

Compliance with Ethical Standards

Conflict of interest: The authors declare that they have no conflict of interest.

Ethical approval: All procedures performed in this study involving human participants were in accordance with the ethical standards of IPMA Serbia and with the 1964 Helsinki declaration and its later amendments or comparable ethical standards, which was confirmed in the Permission No. 8-18 issued by IPMA Serbia Ethical Committee.

Informed consent: An informed consent was obtained from all individual participants included in the research and the data used in this study is completely anonymized.

References

1. Abbasi GY, Al-Mharmah H (2000) Project management practice by the public sector in a developing country. Int J of Proj Manag 18(2):105–109
2. Arrow K, Bolin B, Costanza R, Dasgupta P, Folke C, Holling CS, Jansson BO, Levin S, Mäler KG, Perrings C, Pimentel D (1995) Economic growth, carrying capacity, and the environment. Ecol Econ 15(2):91–95
3. Barber M, Jackson S, Shellberg J, Sinnamon V (2014) Working Knowledge: characterising collective indigenous, scientific, and local knowledge about the ecology, hydrology and geomorphology of Oriners Station, Cape York Peninsula, Australia. Rangeland J 36(1):53–66
4. Beckerman W (1992) Economic growth and the environment: whose growth? Whose environment? World Dev 20(4):481–496
5. Brooks J, Waylen KA, Mulder MB (2013) Assessing community-based conservation projects: a systematic review and multilevel analysis of attitudinal, behavioral, ecological, and economic outcomes. Environ Evid 2(1):2
6. Bulkeley H (2005) Reconfiguring environmental governance: towards a politics of scales and networks. Polit Geogr 24(8):875–902
7. Feiock RC, Stream C (2001) Environmental protection versus economic development: a false trade-off? Public Admin Rev 61(3):313–321
8. Hall M, Holt R (2002) UK public sector project management—a cultural perspective. Public Perform Manage 25(3):298–312
9. Hart D (2018) When does environmental regulation stimulate technological innovation? Information technology and innovation foundation. http://www2.itif.org/2018-environmental-regulation-innovation.pdf. Accessed 25 Feb 2019

10. Igbokwe-Ibeto CJ (2012) Issues and challenges in local government project monitoring and evaluation in Nigeria: The way forward. Eur Sci J ESJ 8(18)
11. Istrate L, Marian L, Ferencz IS (2014) The use of a project management application in managing investments in a public administration institution. Proc Econ Financ 15:1732–1739
12. Jałocha B, Krane HP, Ekambaram A, Prawelska-Skrzypek G (2014) Key competences of public sector project managers. Proc Soc Behav 119:247–256
13. Kozhovska E (2018) Integration of health, safety and environmental principles into industrial project management. Eur Proj Manage J 8(1):33–39
14. Obradović V (2018) Contemporary trends in the public sector project management. Eur Proj Manage J 8(2):52–56
15. Petrović N, Bošnjak I, Nedeljković S (2017) Disaster risk reduction for sustainable development goals. Eur Proj Manage J 7(2):27–35
16. Rwelamila PMD (2007) Project management competence in public sector infrastructure organisations. Constr Manage Econ 25(1):55–66
17. Spalek S (2014) The influence of country of origin on project management: an international empirical study. Proc Soc Behav Sci 156:4–7
18. Santos V, Varajão J (2015) PMO as a key ingredient of public sector projects' success—position paper. In: Procedia computer science, vol 64. Elsevier B.V, pp 1190–1199
19. Savić Z, Bogetić Z, Dobrota M, Petrović N (2016) A Multivariate approach in measurement of the sustainable development of European Countries. Manage J Sustain Bus Manage Solutions Emerg Econ 21(78):73–86
20. Stumpf M, Brandstätter M, Borghoff J (2019) Sustainable project management and the role of integrated communications (IC). Eur Proj Manage J 9(1):10–20
21. Thompson MC, Baruah M, Carr ER (2011) Seeing REDD+ as a project of environmental governance. Environ Sci Policy 14(2):100–110
22. Todorović M, Obradović V (2018) Sustainability in project management: a project manager's perspective. Sustain Growth Dev Small Open Economies 88
23. Wirick D (2009) Public-sector project management: meeting the challenges and achieving results. Wiley, Hoboken, New Jersey
24. Yanwen W (2012) The study on complex project management in developing countries. Phys Proc 25:1547–1552

Part II
Research on Integrating Sustainability into Project Management in Different Industries and Regions

Chapter 7
Integrative Intelligence: A Great Deal for Sustainable Rural Development

Jesus Martinez Almela, Jorge Flores, and Jorge J. Peart Mijangos

Abstract Possibilities of human development in rural represent a complex research topic that shouldn't be broken; neither theoretical-explicative perspective nor pragmatic one. Today more than ever, interconnectedness and depth of human problems forces even serious and empowered entities to venture new approaches; radical ones that, until recently, would have seemed almost extravagant addressing questions like "How could sustainable development… have any chance of even starting and of consolidating in the long term, unless it was based on 'sustainable thinking', a 'sustainable ethic' and 'sustainable cultures". According to this mentality, only by asking the right questions can we identify appropriate solutions, if there is one, before it is too late. According to long-range or long-term vision, the future is not written in the nature of things, but takes shape through the sum of human acts. In view of the current hegemony of the short term, and which submits to the tyranny of the urgent at the expense of what is important—as usually happens in the world of finance, the media and politics we can only have foundations reasonable for hope if we begin to rehabilitate the idea of "duration", the concept of long-term project in another word "sustainability".

Keywords Competences · Extensionism · Project management · Smart rural · Sustainability

J. Martinez Almela (✉) · J. Flores · J. J. Peart Mijangos
Bioagroprojects Biotech PPM, C/Bisbal s/n, P.O. Box nbr. 285.12540, Villareal, CS, Spain
e-mail: jma@bioagroprojects.com

J. Flores
e-mail: erato739@gmail.com

J. J. Peart Mijangos
e-mail: jorge.peart@ipmamexico.com

J. Martinez Almela
IPMA Mexico, C/Carlos Pereyra 6-97, 08200 Iztacalco, Ciudad de Mexico, Mexico

R. Cuevas et al. (eds.), *Research on Project, Programme and Portfolio Management*, Lecture Notes in Management and Industrial Engineering,
https://doi.org/10.1007/978-3-030-60139-3_7

7.1 Introduction

Smart Rural (SR) bets fundamentally on the competences to understand in a prospective way the key interrelations of the rural universe in the world, and from there to be able to act before its inherent complexity, with its problems and its opportunities both major. Probably no scope is as encompassing and decisive as this and on which it depends, mainly, from the food of humanity to the preservation of the Biosphere. For this SIG, SR implies above all an integrating intelligence in rural programmes ant their projects, and that necessarily will have to learn to take a position in a universe that encompasses between its thousand and one dimensions (only referring to its immense spectrum of real determinations: from policies and agrarian laws, up to climatic and atmospheric regimes; from culture, territoriality and peasant identities, to the debacle of a progressive loss of the potential of agricultural soils throughout the planet; from agroecosystems in the tradition-modernity transition, to the potentials and contradictions inherent in cutting-edge technologies—ICTs, GMOs, satellite monitoring, alternative energies, etc.; in short, from markets, competitiveness and total globalization, to social justice, development, quality of life and human dignity. To begin to recognize them, all of them, imbricated in rurality in the world, is part of Smart Rural effort: to understand integrally, to articulate potentials; project to transform and also to preserve, because today as never before, understand the connections that exist between things and relate ideas in a way that acquire coherence and meaning become vital answers.

The Mexican historical experience—with its undeniable mistakes, achievements and discoveries [2] makes the former considerations inescapable, complexity and interconnections that must be recognized and taken advantage of. With its immense ethnical and cultural variability, along with its exceptional geographic and ecological diversity, the techno-economic potentials of rural production in Mexico are still magnitudes we are just starting to glimpse and measure in all projected initiatives for a real and desirable change and transformation [2]. In other words, only under the condition that we can understand the subtle but decisive interrelationship between forces, factors and trends apparently dissimilar and independent, then can and should actions projected upon "the rural" be an opportunity to strengthen processes as urgent as equally co-determined. Such would be the case for interrelated achievements such as fair trade, the creation of hope for populations in rural environments, inclusion, gender equality [concomitant to] handling the farming and forest environment and food production [7].

Today we must be capable of understanding and projecting the rural universe as a real source of global solutions, in our present and imminent future. These solutions include a universe of potentialities from where we can restore peace and social well-being, food security and health for everyone, participation, equity and solidarity and political-economic justice. Furthermore, it is a privileged universe from which we can fight poverty and co-relationally the devastation of ecosystems at a planetary scale. Legitimate but ambitious aspiration of Smart Rural initiative from SIG-Special Interest Group Mexico from which we can evaluate and globally share national experiences waiting to be capitalized upon; i.e. identifying and systematizing better practices at a global level on projects focused towards sustainable development; facilitating the exchange of experiences in regard to smart management models for rural development. A horizon and perspective from we can contemplate contributing to the construction of a global interest, exchange and support network, so that the PM-project management profession by competences may achieve a greater and better world positioning in this priority problematic for the future of humanity. Smart PM focus as alternative initiative capable of reconciling, in their design and proposals. The search for developing models for innovation and territorial competitiveness management based on local strategic programmes and projects, through utilization—among other tools—of information and communication technologies so as to facilitate the exchange and management of knowledge, globally and in real time.

7.2 Rural Mexico: Gross Lines Towards Project Management Insight

From any right "diagnosis" it is evident that "rural Mexico's" stand out by themselves as a specific territory with a significant population that posted challenges and opportunities, clearly differentiated from the all-urban contexts, their own problems, logic and dimensions. It recalls the fact that rural areas comprise close to 90% of Mexican territory but, further on, its significant heterogeneity from territorial perspective; specific questioning of what is the actual value that states, regions and Mexico as a whole accrue from its vast universe which means that, from Smart Rural projects vision, should not necessarily pursue the same agro productive objectives, and use the same tools in every state or region. Social sphere, for example, involves poverty alleviation, education and health policies (therefore how they have acquired an increasingly relevant role in rural policies). The true and liable diversification of the rural economy will require align projects to perform accordingly. Several of the opportunities and challenges of the rural territories implicate the involvement of different stakeholders that might not have had an original or essential rural perspective, where strong competent leadership is important for the coordination of efforts of multiple agencies.

Ambitious horizon, effectively, this assumes that we are capable of understanding and projecting the rural universe as a real source of global solutions, for our present and for our imminent future. These solutions include a universe of potentialities from where we can restore peace and social well-being, food security and health for everyone, participation, equity and solidarity, and political-economic justice. Yes, it is a privileged universe from which we can fight poverty and co-relationally the devastation of ecosystems at a planetary scale.

The property regime of 85% of the forest land is communal land (ejidos), and this particular characteristic makes rational control difficult since there is no clear responsibility for the good or bad use of these assets.

Mexico consumes about 80 km3 of water annually, near 17% of the natural availability (68% from superficial and 32% underground). The most predominant use is agriculture, 83%; public use 12% and industrial 5%. Probably the average efficiency of the use of irrigated water is 46%. For the Center, North and Northwest regions, the level of use is 47% of the natural availability, level of "high pressure". Irrigated water lands have increased from 750 thousand hectares, in 1926, to 9 million hectares around 2007 [1].

Production of basic grains has remained stalled due to a reduction in the harvested area affected by adverse climate, a decline in the producer prices, increasing costs of supplies and inefficient scale of production.

Today, the limits of agriculture are set by the fact that of the 198 million hectares of Mexico, only 11% are arable territory and most of it is already with that use. Unfortunately only large producers, in general, have scale and access to financing and technology, which are the major drivers of productivity increases and improvements [8].

There are many rural Mexico's. Generalizing into "rural" therefore might be an interpretative mistake, since "the great challenges of certain rural regions might completely blur relevant opportunities for other rural regions". Following OECD strategic characterization we can here identify at least about these different types of Rural Mexico:

7.2.1 The Rural Poor

Dispersed, more or less inaccessible; low levels of public services, almost isolated from markets and thus characterized by low productivity, subsistence agriculture. This segment still now has not been able to obtain economic value of the natural richness. These rural areas propitiate land use changes and expanded the agrarian frontier at the costs of forests, mainly during the past four decades, without significantly improving the poverty condition of its inhabitants. The dispersed character of these localities is one of the major obstacles to overcome in order to open them a variety of opportunities that they don't have at reach. Improve accessibility is a determinant of diversification of income and higher salaries, thus the rural roads and communication network are essential. Continuing with the approach of creating critical mass of infrastructure and services is crucial. As well as continue covering the deficit in basic public services: living in a "solid floor" house, with electricity and access to potable water and sanitary services really marks a difference in the basic conditions of the rural population. Bringing key stakeholders to this effort is a major future challenge.

7.2.2 Rural of Intensive Agriculture

In contrast to the previous one, has had significant productivity gains because its orientation towards and competitive in export markets but depends highly on irrigation; mainly contributes to the depletion of water reserves, however that land represents only 29% of agricultural land, where 55% of national agricultural output is produced and the origin of 77% of agricultural exports.

7.2.3 The Rural Diversified

It has adopted the "economic vocation" of the region (see 'New Rurality' below), as well as the nearby cities and become location for industrial parks and/or services taking advantage of lower costs. Thus opportunities for economic development are thus more and more linked to non-farm activities, because agriculture itself has had an important transition in the past 15 years. There, large natural resources potential could generate substantial economic value if exploited in a sustainable way and taking advantage through better-organized forestry and fishing-aquiculture activities. As well as, there is a tourism potential, rich in natural, landscape and cultural amenities, close to centers of massive attraction where communications and training can bring an alternative source of income to local inhabitants.

7.2.4 The Rural Peri-Urban

Its life and dynamic is closely integrated to urban areas, with two extremes: the "poverty belts" generated by migration from rural areas that require extending poverty alleviation policies to urban contexts. On the other, there are areas that have also become home for accommodated people seeking a calm refuge to live out of the big cities, leading to increasing commuting flows to and from urban areas, and a growing demand for services.

Matter the fact, particularly Mexico, paradoxically contrasts with available resources and basic life quality level in rurality. Mexico occupies one of the first places among mega-diverse countries of the world (fourth place). It possesses an enormous multiculturality in its rural population and a substantial part of such an ecological wealth is in the hands of indigenous groups. In Mexico, for example, rural producers possess 80% of forests and jungles and 74% of biodiversity, according to CONABIO and also CDI. Likewise, social rural property (common land and communities) is responsible for almost half the total agrarian production. Out of the more than 31.500 agrarian centers in the country, 15.584 are located in forest and jungle zones, with a surface close to 63 million hectares (Mexico is today a world leader in forest surface certified in community forest handling). Nevertheless policy, incentives and institutions have not yet been aligned with conservation and sustainable economic benefit from these resources. The growth of the agricultural boundaries, illegal fallings forest fires and water depletion are the key challenges for sustainability in rural areas. Rural areas with potential development of rural tourism and renewable energy as complementary activities will help looking for sustainability and integration as well.

Mexican "Rural" universe comprises more than 85% of its territory when "rural territories" means the basic unit of assessment addressing further interventions with a comprehensive understanding about demography and social dynamics recognizing the heterogeneity within rural areas. Since 2000 rural population has been declining due to migration of youth people with a high economic and socio-cultural impact in more than 24 million people living in these areas.

Land property regime since the Agrarian Reform following the Mexican Revolution (1910–1917), tied many people and their descendants to atomized plots of land an indirect cause of the attachment of population to live in isolated rural areas. Matter of fact, Mexican approaches to rural policy and intervention in general during most part of the twentieth century have based on three main pillars: the agrarian reform and the communal property of land, the prevalence of an agricultural focus, and the vision that rural economy was in a pre-market condition.

Accordance with Farming, Fishing and Food Development Sector Program 2013–2018 (PSDAPyA, Spanish acronym), in the "Challenges and Opportunities" section, greatly emphasizes the topic of smallholdings and their low production scale in terms of "Close to 80% of those who produce within the countryside have plots of land smaller than five hectares, which implies they don't have a production scale and that they present organization problems that greatly limit their productivity and competitiveness [...] the lack of scale economies generate high production costs (seeds, agricultural chemicals, machinery and equipment, fuel, among other inputs), in addition to generally not being destined to produce high value products and at production volumes that are small for the size of the production units". Even though some 2.7 million producers in Mexico do have an interest in introducing their production into the markets (whether local, national or in some cases even, international). The most worrisome part of this second group of farmers is that they still have important limitations or restrictions for an adequate commercial participation, due mainly to:

1. Low production capabilities: whether due to lack of technology or lack of knowledge. This is also because of the severe ecosystem deterioration of their main production resource: the land;
2. Because in many cases they have shown a limited capacity or willingness to organize themselves, meaning, to agree effectively and lastingly; to organize to buy together, produce more and better and therefore sell their products at higher prices.
3. Because they have little access to a very important factor: training, which is always necessary to cover their learning needs. They also have little access to

technical assistance to help them identify and solve technical, management, sale or commercialization problems (among the most important ones). Many technicians work solely on "getting support", instead of working on their true role: training and educating, helping solve problems, technical and productive, labour organization, sale or commercialization ones, etc.

In this case, the population is isolated from markets and public services and structurally attached to subsistence agriculture, in one word, poor. Currently there are approximately 5-five million rural production units (UPRs Spanish acronym) located in primary production (98% of the total), nevertheless these entities occupy links in the production chain (basically primary production), where predominantly they only retain 10% of the produced wealth in the array of production chains-value networks. In other words, producers buy alone, produce alone and sell alone; they buy expensively, produce expensively and sell cheaply as well as though there are innumerable resources and production potentials in public land and communities, that is to say, a vast amount and diversity of natural resources in social property, generally they are not a source of wealth through use and sustainable handling: they constitute rich public land with poor shareholders. They are an atomized scenario in smallholdings, without economic association and without scales to access the market and despite there is available technology, it isn't appropriate and sufficiently accessible to producers and technical assistance is not enough.

It is clear today that this situation is the root of the structural imbalances in rural economy—with national structural dimensions. From Mexican experience at least, to be intelligent face to sustainable rural programmes and their projects suppose no one would question that a harmonic and efficient development there is a consubstantial for our societies managing to overcome the regional imbalances that threaten governance and social peace, political, economic and institutional stability. So, with "structural imbalances" in Mexican rural economy means:

1. It generated the lack of profitability in the immense majority of rural production units (UPRs) and their decapitalization;
2. It fosters a high commercial brokering in raw materials, inputs and products;
3. It feedbacks and exacerbates the negative aspects of smallholdings;
4. It generates and perpetuates rural population migration processes;
5. It encourages a lack of generational replacement and ageing and feminization of the rural population;
6. It implies high investment and credit risks, because of the inherent and contextual risks in primary production;
7. Stresses regional backwardness;
8. It limits value addition, the presence of other economic sectors and the development of the territory's internal market.

These are substantial causes that strengthen and power each other for the conditions of regional backwardness and rural poverty.

This reality underlines the fact that even though rural areas in Mexico confront similar issues that other countries face, namely, lower productivity and income, out migration, deficient public service delivery, etc., the issue of rural development is a major priority and challenge for Mexico. Tackling effectively the problem through appropriate policies is a matter not only of economic transcendence but mainly of equity, having substantial implications in terms of social cohesion.

7.2.5 *Rurality, Intermediate and Urban*

Rural and rurality meaning is based on territories with low population and no urban settlement. Intermediate spaces are not classified as such sometimes referred as "Villages", depending on the population living in communities into the rurality, and highest human settlement means a context predominantly defined as urban.

Anyhow the importance now bestowed to organization and educative challenges for the rural population; so, main strategic lines of action must be according as National Development Plan-PND 2013–2018:

1. Promoting in rural environment inhabitants production capabilities with entrepreneurial vision, exploring alternatives so they may incorporate into the economy more productively, from the standpoint of better production and creative capabilities for young, female and small producers.
2. Advancing commercial politics focused on farming businesses and planning for the supply demand balance; based on timely supply and at competitive prices, key points for getting food security for the country.
3. Emphasis on sustainable production projects, on development of technical, production and commercial capabilities, such as integration of local production and commercialization and investment circuits.

Approach places so much interest in the country-city relationship, needing even to "blur" this very sharp distinction. One only needs to think of how rural and urban job markets have had an ever-increasing integration, or even, the fact that country households obtain important extra-agricultural income in cities and rural populations as an also growing trend. Following the example of "New Rurality" school in Mexico, SR-Smart Rural recognizes the reassessment of rural spaces as a strategic continuum with the urban: "…that the rural is a scattered population focused on the farming sector, then going towards the reconstruction of the work and political objective by defining the rural environment as the territory constructed through multiculturalism, integration, diversity and economy can establishes a new functional rural-urban integration relationships. This redefinition of rurality implies a new concept of planning… and it supports it in the need to build a true rural citizenry. As well as, New Rurality "opens fertile ground for the incorporation of concrete strategies for implementing institutional reforms more consistent with the social, economic and political changes and energies which have conditioned the possibility of an effective development. [10]

The above also allows us to highlight the importance of local areas in which systemic economies are generated for a more or less diversified array of business. In general they achieve an expansion of activities located in the environment through linkages between business that can also encompass anything from support systems for primary production of intermediate supplies, up to benefit, transformation and final products [3].

These business, as enterprises, involve key local economic meanings, promoting development with a territorial focus in this sense: they constitute economic functions; they are carried out through economic agents; they are performed by businesses. As well as, they imply profitable economic scales and value addition processes, since they constitute a basis for the internal market [9].

The division rural-urban is deep, significant in terms of standards of living, among others. In Mexico 26 of its population is gathered in 3 metropolitan cities, and 23% of its population lives in localities of less than 2,500 inhabitants. In the rural milieu, GDP per capita is less than half in rural areas than in urban areas, and extreme poverty is mainly a rural phenomenon (60.6 per cent of the population in extreme poverty lived in rural areas). Even if Mexico can be among the highest 10 economies in the world in terms of GDP (77% produced in urban areas) it score low in most average indicators, from GDP per capita to educational performance and human development tests. According to the CONEVAL (National Council for Social Development Policy Assessment) within the framework of Public Policy on Poverty and based on indicators generated by INEGI (Population and Housing Count), out of the more than 24 million of rural inhabitants more than 21 are considered to live under poverty conditions, "in food, in capabilities and in property", in adherence to its own characterization. (This can be consulted in the electronic webpage for the Council's document *"Construyendo un sistema de planeación y evaluación de la política pública"* [Constructing a Planning and Assessment System for Public Policy].

Between 2000 and 2005, however, several factors, still for surveying and understanding, have contributed to closing the gap between urban and rural income, with a significant reduction in extreme poverty levels, about 32% approx. The evolution from agricultural to other non-agricultural activities in rural context today represents more than 50% of income even in dispersed rural areas.

In fact the availability of non-agricultural activities will contribute to the improvement of quality of live and more equal society in rural areas, but still more qualified and formal jobs are needed looking to create hope kipping rural population on-site. Rather, we must consider too what happens with this supposed reduction, at least ever since the nineties not just in Mexico but symptomatically in other Latin American countries, is the reduction in percentage of total poverty in the rural sector is not due in general terms to a successful reduction of the incidence of rural poverty but rather to rural-urban migration.

7.3 The Rural Universe from Project Management: The Source of Solutions for Our Global Future

From the Mexican experience at least, today we can better understand that availability and supply of natural resources and their production potential in the rural environment is not only an economic and subsistence criteria. Likewise, it decides patterns of appropriation, permanence and territory identity for human groups—their very condition of possibility. Thus, this is about no less than civil construction (and reconstruction) historical processes, as warned by some experts in the Mexican and Latin American experience in general. A real new rurality will be because influences the capacity to fulfil important functions linked towards conservation of natural resources and construction of social capital including the social and political function of communities.

Accord to SR we believe that nothing would be more misleading in realistic rural development—in our present world– than presenting "recipes", common place or supposed procedures and pre-established prescriptions. For instance, development experiences and learning about the Mexican agricultural sector. From actions with key actors—such as rural extensionism and country organization leaders—one can identify and strategically encourage projects that relate country potential and needs with technological innovation incorporated within the projects. IPMA México has confirmed the exceptional potential of said human capital through training and certification, to the present date, of almost 1700 rural extensionism. This is through the process of promoting the professionalization of Project Management under its global standard ICB, but implied in regional development actions. Thus it has been possible, to a large extent, to line up conditions, actors and key scenarios. That is, this relates viable integrating projects, guided by agricultural specialists certified in the IPMA international standard. It considers compatible commercial, technological innovations and financial options, and from this considers to the possibilities for inducing transformation and improvement in the objective life conditions of general country and rural populations.

7.3.1 *About "Extensionism"*

The main value and sense of scientific knowledge as agronomical research and education lie in its availability for all producers in agro. So, as "universal asset", these kinds of knowledge demand to be "extended" (to enlarge, to spread out) to ultimate needs in agriculture; a bridge to improvements in productivity and quality of rural life by means of extension of the agronomic science born in academy and laboratories (about soils, seeds, technologies, fertilization, resources, biology of plants and animals, etc.). Just that was the founder philosophy to create in Mexico in 1933 the "Departamento de Fomento Agrícola" (Farming Promotion Agency); later Extension Farming Agency (or Departament), around 1950. From since, "extensionismo" (in Spanish), meaning "extensionism" (in English) as founder principle, as thinking school of development, and as real national endeavor has lived a history of contingence and complexity. A great potential still life, no doubt, despite of critical episode and marasmus neo-liberal, with contraction of Mexican Govern and their responsibilities and intervention at all stages of commodity distribution channels, price supports, barriers to imports and subsidies in inputs such as credit, irrigation, energy, fertilizers and, of course, technical assistance and transference. The word "agricultural extension" it has different meanings depending of the context, region and nation as well., a.e. in the United States Agricultural Extensionism is a common tool managing grants and help through annual programmes like Farm Bill (cooperative extension service).

Indeed, some of the main essential instruments for this vision and strategy will encompass everything from capital, credit and financial services, to supports for infrastructure and technification, said technological solutions, among others. However, it is impossible to conceive any project worthy of this name without taking into account factors—mostly decisive—such as solutions and innovations in the competence development of each implicated sector (through learning and knowledge in accordance to each enterprise and business). That is without mentioning the organizational, contextual, planning and communication competences that every duly projected change demands. In consequence, is a basic and smart premise that producers are capable of identifying central problems they must focus on to improve their context and motivate solutions that allow for the generation of development. In consequence, one must assume that the systematic and participatory exercise of examining, criticizing and rectifying their own knowledge of their production environments and realities is a basic need for reaching information with strategic value for their organization.

This information is ordered and put into a hierarchy so as to identify and encourage their projects. That is, those that because of their characteristics and potential demand and also cause a strengthening of production organizations. They also generate the conditions for the most pertinent and appropriate local development in each of the regions they are inserted into. Remarkable learning here: neither the techniques nor the tools, nor even the methodologies (all the media), may, by themselves, bestow competences upon the individuals. These last ones are the learned capacities for subjects—with their know-how, skills, dispositions and attitudes—the ones that will give their due value and potential to the instrumental media. Therefore, education was there an end unto itself, that is, education that builds significant learning. It made people more competent; with competences that they could understand, develop and, above all, apply to our efforts and endeavors for transformation and improvements.

According to this logic and mood, training-certification programmes have been encouraged for leaders and professionals as well. This is in regard to social and production transformation within the local environment, so as to contribute to expanding competences for strategic analysis of their regional reality and local context and to their insertion in economic national and international dynamics. From here we see how important it is to strengthen the role they must play when designing strategies for transforming their reality, as a direct incidence in the significant improvement of quality of life in rural communities. This is how we emphasize the importance of their competences in organizational processes, in technological developments implicated in each viable market perspective and, in general, in the direction and management of development projects with a focus on territory, since strategic concept of territoriality implies a whole web of ideas and notions that are very important in the study and project of sustainable development. These notions, today, must come from areas as profound as ecology and biology, economy, anthropology or political geography, among others. We could consider, for example, the implication of sustainability in human development; with a focus on territory, this could also be considered within the behavioural and symbolic spheres as well as the historical and cultural ones that humanize the complexity of territories, with their all technological innovations, natural resources and economic production potentials.

Rather than a mere neutral and naïve instrumentalism, SR has assumed an imminent need: to configure and encourage new profiles in rural development project managers. This is without underestimating the great potential entailed by ethical and strategically instrumented technology, on the contrary. We refer to functional profiles based on a holistic and human conception of PM. It should be capable of integrating, in organic standards, capacities (strictly competences) as vital for all human needs as are, among others: Design, Planning and Project Control and its Stakeholders. Actually, they are but mutually and organically enhanced through other essential capacities: in Strategy, in Power and Interest or in Culture and Values. Or rather, they are thus through the capabilities of Negotiation, Self-reflection, Ingenuity and an Attitude, of course, towards Ethics, in Appreciation of values and Personal Integrity [5]. With the evolution of its competence regulations, IPMA attributes its originality to an organic integration—the whole is more than the sum of its parts—of performance standards for a new generation. They are available today for their realization or better yet "performance" (in the sense of creative interpretation) before each project as a microcosm of possibilities and challenges.

Encoding the different facets of the competition in the PM is not only a condition to advance in the professionalization of the discipline; also, it is the way to scale up its position within the most pressing scenarios in the major contemporary problems: food, health, environment, education, human well-being. And precisely all these areas of priority interest, and many more, include the rural universe. It is important to advance in the systematization and codification of the competences that could be the most decisive in comparison with the practices, scenarios and human dynamics of rurality. This is a determining issue and one that hopes to be duly grounded and exploited in presumed Smart Projects in rural development. We must consider that peasant organizations and producers of the sector really constitute "special" entities. As cells of a territorial fabric, due to their characteristics, needs and potentials—from poverty and lag in many cases, rooting or migration, to cultural determinations, tradition, values and identity—they often constitute challenges that are only partially understood and, objectively, distinguishes them and departs from the functional logic that can characterize the industrial or service sectors of urban environments and their governance contexts, their organizational structures and therefore the degree of formalization that is possible in their standards, laws and regulations with who manage to normalize their varied operations and practices.

The interest in understanding and enhancing the capacities of individuals in the vital interaction with their environment is focal for the area of Competences in Perspective that is, trying to confront the environments that dimension those nebulas of reasons and motives that underlie individually, collectively and socially when undertaking and paying for projects. We know that especially the Programmes come into existence to achieve strategic objectives, and precisely, are rural development and entrepreneurship initiatives that should be understood and promoted from this spectrum of scope of action not only to allow the implementation of changes and obtaining of benefits, but in the substantive terms of what, why and for what of the interventions projected within a set of scenarios that involve the corresponding populations, in the particular phenomenon of their productive and techno-economic practices in general, their social and demographic dynamics, political-ideological adhesions, mixture of values and beliefs that are often rationally inscrutable, etc.

We insist that especially in this area—those of human development in the rural world—PM inevitably transcends performance parameters and technical, organizational and administrative functions typical of temporary and limited initiatives, reaching to imply or induce changes in life and actions of real communities; by influencing decisions such as investing multiple resources, or financial responsibilities and market risks that carry the expectations of a viable peasant entrepreneurship, for example, inseparable also from political commitments and from the "games of power", which is never dissociable from the spheres of ethics and social responsibility in the face of the change and transformation to be sought over the living conditions of the people involved, this in the direct realization of your projects. Change and transformation are needed through projects and their programmes but the disruption created by individuals around project and programme management need new competences, new attributes such multicultural, skill and abilities applying lessons learned from the recent past.

7.4 Project Management Facing Mexican Rural Context: Some Opportunities

From PM rationality, and very important into Mexican rural context, one might considerer that, due to its maturity, amplitude and realism in a kind of pro positive capacity, a decisive moment in the face of rural problems in Mexico is represented by the "*Seminario Internacional sobre Desarrollo Rural y el Sector Agroalimentario*" [International Seminar on Rural Development and the Agrofood Sector]; with plantations fully in force since 2006, the year in which it was carried out. It is worth mentioning the following priority action proposals:

- The deepening of decentralization and federalization actions, to provide local initiatives with greater instruments of promotion and support for the development of the field.
- The creation of planning instances with the joint participation of government and citizens.
- In the process of decentralization, it is important to develop administrative systems and technical capacity in the formulation of projects for local government and to increase democratic construction to avoid local "caudillismos" (Spanish word meaning "warlordism" a kind of autocratic leadership) replacing central verticality.
- The functioning of the Councils of Sustainable Rural Development, to promote spaces for social participation and construction of a new rural citizenship.
- The Law of Sustainable Rural Development has opened spaces for social capital as the element that seizes the strategies of collective action with decision-making power and territorial approach.
- The Committees "Product System" are conceived within a new institutional framework, around the orientation and allocation of investments to improve the competitiveness of agro-food chains, with the participation and representation of their main individuals so that they assume responsibility in the integration and cooperation with reliability, competitiveness and equity.

7.4.1 The Project Manager as Agent or "Extensionism" for Sustainable Rural Development

In this sector the profession or discipline of the project management, strictu sensu, has a still scarce or incipient presence. Not to mention the degree of clarity about the essential competencies for this special field of action. It is urgent to correct this limited presence of the discipline, still underutilized: more innovative and innovative projects, designed and executed simultaneously through programmes and portfolios; thus, in the interest to promote the role of Project Managers, in harmony with the best meaning of the specialized services to the rural world, as agent of development in the sector. Within specialized services within rural areas in Mexico, important deficiencies have been identified, deficiencies that have not been resolved between instances, forces and decisive factors (in synergy) for the sector. Such is the case of the disengagement that they exhibit, nothing less, potentials such as those of research and academia with the final producers and their real problems; this objective situation, that of not being able to reach them, and that, in varying degrees and for various reasons, limits the level of coverage and "impact" in technology transfer. There are analyzes that can tell us much more about the true background and meaning of the phenomenon of untying and of so many underutilized potentials; especially in view of the urgency of dimensioning feasible solutions to "root" problems, such as those of food, poverty, natural resources or the promotion of profitable economic units.

Today we know that, when considering some of the most important limitations of professional or specialized support to rural producers and their real needs, in general there is clarity and consensus among various critical approaches in this regard. Thinking about PM competences (overall Perspective and Persons), and according to authoritative recommendations, the role and function of various development agents involved in rural projects can be characterized from aspects and inertia in their professionalization such as

- Deficient systematization between theory and practice
- Disperse thematic subjects
- Little connection to community and productive problems
- Low execution capabilities, in general
- High value of material support and reduced value for intellectual type (effective learning)
- Labour orientation towards the government supply, ergo with little capacity to undertake
- Reduced training in diagnosis and planning
- Little training in agricultural businesses
- A lot of study on poverty but "we do not teach to generate wealth"
- Poor social participation and reduced leadership
- Priority to the management of government funds and some productive assets, although lacking frameworks for empowerment or induction
- Delivery of support to "beneficiaries", but without accompaniments and monitoring of impacts
- Lack of training and education programmes for producers, focused on real business opportunities
- Little or no ability to "know-how to do developing the capabilities of people"

Nevertheless, from Practice concerns, the former, must be too assessed from an evolutionary perspective—its potentials—since new individual competence development and assessment through organizational competence building a different approach managing project portfolio and its level of maturity that will help individuals and organizations making things happen.

This characterization today is predominant in this, however, key vector of the Mexican rural area, and represented by diverse specialized agents of development; notwithstanding these functions (assistance, training and management in projects of producers) from could and should be controlled "the application of resources and execution of work plans, accompany the development of projects [rationalize] the use and management of natural resources, improve marketing and revenue, and increase productivity". The role of the adviser, extension agent ("extensionism") or in general of the development agent in the end is more important than any material support by themselves—says the quoted specialist—and above all, to think that: "the only investment that really transforms people", since it transcends even the only productive aspect. Due to the delicate nature of the role of this type of specialists or professionals in sectoral projects for development, and therefore their training needs, they should be subject to more and more sophisticated methods of evaluation and certification, note other authoritative voices of the sector: "From standards defined with specialists and institutions of higher education and research. Generic and specialized standards for the different productive activities of the rural environment [and its] different regions" Its professionalization will necessarily imply to base and normalize competences, from regions and crops, to productive practices, organizational and learning processes, especially among small and medium producers, just to mention some priority areas of competence. Since this reality and challenge, Project Management based on competences is able to engage its main potentials.

The need to learn will be always proportionate to the size of their goals. That is why experts insist that every organization interested in integrating project-based culture needs to define its goals in order to develop the related organizational competences and manage the individual competences of its people, and thus foster the development of their competences as a whole, and to do so, it needs new competences by domain and role level to be achieved; according to the needs of the given business and the competitive environment in which it operates

Because, today, rural human development by means of new alternatives geared towards local-territorial action, "...must be sustained by the conviction of its actors, given that it becomes necessary to strengthen the level of knowledge and awareness, in regard to development planning [just where] we require a nucleus of technical actors who can drive, guide and facilitate regional action together with a political and strategic sense [4].

So, thinking about Mexican rural great challenges as DP opportunities—and its realistic and lasting endeavors—from the experience and studies by main specialists there is no substitute for "the capacity of a local society to formulate collective progress purposes for the material, equality, justice and sustainability, and to mobilize local endogenous resources needed for their obtainment" However they consider Mexican society, especially phases like the current one, to be in conditions difficult for the collective conception, construction and operation of a social-territorial project and it is crucial to overcome this situation.

It would seem the best opportunities are being created by other forms of organization and relation between the public and the social, through associations, institutes, networks and civil society organizations, that ever since the mid-nineties at least have been stimulating proposals with voluntary participation of public and private actors and from civil society [...] If we intend to count on policies and institutions that promote territorial economic development in Mexico, we must be open to substantial changes sustained by diversity [6].

A great deal to rural integrative intelligence from projects and its direction: to impel endogenous conditions for human development, therefore, the competences and learning involved in each endeavor. Thus, to acquire the learning needed to embark in a new company or business will provide the development subject with its main strength: applicable competences. Even, today the best and most modern approaches to public development policies confirm that, for example: "In the 21st century it has been shown that the weight of knowledge in productivity increase is higher than that of other factors such as land, capital and labour [...] Knowledge [is] the central axis of every change strategy in rural areas, which implies renewing it and applying it to achieve effective changes in the living conditions of rural society [11].

7.5 Devising Tools for Rural Development: SR Against Naïve (and Danger) Narrow-Instrumentalism

Smart Rural approach agrees that the sustainable rural development is a multifactor and multidimensional programme, a. e. long-term, long-winded, inherently difficult and prolonged, and it can only be understood through a trans-generational perspective: the idea of rural development as a condition of better environments and circumstances of life and work for present generations but without compromising the opportunities and rights of future generations. Sustainability is not given in terms of "all or nothing", but instead, it is expressed as levels or magnitudes: "gradients": more sustainable, or less sustainable, in three dimensions indissolubly united and interdetermined (society, economy and environment), where influencing any of the three inevitably moves or displaces the other two, for greater or for less sustainability.

The planned initiative of any productive project in agriculture—planning for feasibility before the rules and demands of the markets as they work—establishes their capacity to generate and retain wealth equitably (which is inseparable from the responsibility of managing to environmental preservation) as a priority. Ultimate commitment, considerer projects for development.

In the economic dimension axis, sustainability in the positive sense fosters the integration of productive chains-value networks based on business plans capable of channelling work organized with associativity and inclusion criteria as a generator of wealth as real community's initiatives. In its negative aspect, i.e. in the absence of the above, it operates to a greater or lesser degree a position of inequality, of disproportionality in the distribution of value generated through human work, leading to isolation, exclusion and poverty.

In the ecological dimension, its positive expression corresponds to the territorial adequacy, which induces the creation of efficient plans for the use, knowledge and management of all kinds of resources, which in their broader meaning includes human work, and which are measurable in terms of the different manifestations of energy. On the negative side, this axis leads to depredation, imbalance, spoliation and depletion of biodiversity and sources of resources, mainly non-renewable resources.

In the positive social dimension, economic integration encourages the participation of human groups in their broader and more varied sense, which, consequently, demands the formation or construction of the collective social subject through competence development plans, i.e. closing gaps between what you know and what you should know, from living work, conscious work and as the means—necessarily—that generate new learnings, goods and services for the peasant and producers organizations, their families and communities. In its negative side, this dimension favours the exclusion, marginalization and contempt of the human being. Indeed, major challenges for any smart project in rural sustainability.

The competent project manager underlies a deep principle against technocratic reductionism: the real potential to empower the people, rural producers and communities as change agents, of its autonomous endeavors. In its most broad but fair sense, "engineering" implies specialized action efficiency through a reasoned, intelligent and explanatory knowledge of different levels of reality to change and intervention (from the molecules in genetic engineering, the markets in financial engineering or machines or agricultural soils in agronomical engineering). Well, a question like the following seems pertinent and timely: What efficiency of action should be expected from "social engineering", especially the one involved in training and in the management of the knowledge and skills needs by producers to develop a given company or project? This is a question that, at some point in transformation efforts—by means of development programmes and projects mainly—will have to be examined and undertaken. Thinking therefore of a "social engineering" of development will involve acknowledging the inevitable complexity of managing people, but also recognizing the relationship of doing this with the training of development subjects and practice communities, i.e. how training (education at and for work), as well as knowledge management, can and should influence the quality of understanding that rural producers need for their development projects before the markets and environments, as territorial conditions of possibility.

Therefore, the most intelligent "device" in rural development would lay in any new machines or technologies? By no means, but social enterprises towards markets by means of cooperative spirit.

7.6 Conclusions and Outlook

Smart Rural strongly believes that the authentic sophistication of current technologies does not lie in improving their intrinsic engineering design, but in the acuity, flexibility and ability (smart) of their human and humanist application: their instrumentation where, when and how they are really needed by people and consequently sustainable as such.

A new knowledge society finds its way into a much more integrated and globalized world, a world which, however, is more perplexed by the fact that the oldest and greatest evils of society, such as war, hunger, poverty, environmental unsustainability and social, ethnic and gender inequality, tend to become permanent problems with which it is necessary to coexist. This implies that today, as never before, it is necessary to use our imagination, creativity, openness and critical spirit in order to face the core problems affecting the development and the destiny of humanity.

We believe that matter implies to go beyond the position based on something as debatable as that common place of the so-called "intelligent" technologies, towards something as unquestionable (and urgent) as the importance of education in technological intelligence; something that would not be limited to the mechanical mirage of mere designs and functions, but would seek to advance towards the human sense of technological applications. Technologies, as means and not ends, have never been or will be intrinsically valuable. Since the real sophistication of a given technology does not depend on its design, as much as it does in its use, application and effects over people and environment life. Therefore, to improve the understanding of the possibilities provided by technologic advance (from agronomic to computer science or even genetic or financial, engineering among others) to the human problems of rural areas, will depend on whether we incorporate these advances upon the basis of projected initiatives where strategy and mission can align with the global commitment of sustainable development. And although the best way to properly apply these technologies is to incorporate them in a methodical and empowered manner into a systemic and systematic contexts called Programmes and Projects, the actor responsible for making it possible should in turn follow a functional profile and a practice philosophy consistent to the times and needs of the current times that we face today: an explicit conception of a new project manager in human sustainability development. The fundamental issue to be defined will be, in the end, the characteristics and capacities required from these new projects professionals, i.e. their competence performance… Unfinished, hard—and smart—challenge to IPMA.

References

1. Alvarado K, Giugale M, Olivier L, Vinh N (2001) a comprehensive development agenda for the New Era. Washington D.C, The World Bank
2. Bindé J (2005) Visión de la próxima generación: América Latina y el Caribe en el umbral del nuevo milenio. [Next Generation's Vision: Latin America and the Caribbean at the New Millennial threshold]. Naciones Unidas—Banco Interamericano de Desarrollo. Icaria Editorial. Barcelona
3. de Janvry A, Sadoulet S (200) Cómo Transformar en un Buen Negocio la Inversión en el Campesinado Pobre: Nuevas Perspectivas de Desarrollo Rural en América Latina. (How to Transform Poor Peasantry Investment into a Good Business: New Perspectives for Rural Development in Latin America) Work document prepared for the Conference on Development in Rural Economy and Poverty Reduction in Latin America on the occasion of the Annual Meeting of the Board of Governors of the Inter-American Development Bank. Mesa Larga-Carmelo, Cartagena, Colombia
4. Ferreira H (2005) "Construir las Regiones" (Constructing Regions) Colegio Mexiquense. México City, Mexico
5. IPMA: IPMA Individual Competence Baselines for Project Management Competence V.3 2005, and the IPMA Individual Competence Baseline for Project, Programme & Portfolio Management. Version 4.0. Zürich, Switzerland, 2005 (v3) and 2015 (v4)
6. López G (2008) Políticas e instituciones para el desarrollo económico territorial. El caso de México. (Policies and Institutions for Territorial Economic Development. The Case of Mexico.) ILPES-CEPAL.p. 160–161. Ciudad de México
7. Martínez Almela J (2018) "Certificación internacional en Dirección de Proyectos para Técnicos y Extensionistas de SAGARPA" (International Certification in Project Management for SAGARPA Technicians and Extensionism). Mexico City, Mexico
8. Mexico Government: "Programa Especial Concurrente para el Desarrollo Rural Sustentable" [Concurrent Special Program for Rural Development]. Poder Ejecutivo Federal. Comisión

Intersecretarial para el Desarrollo Rural Sustentable, Cross Secretariat Commission for Sustainable Rural Development. Ciudad de Mexico (2007)
9. Mosqueda EM, Thennet EI (2019) Herramientas para la formulación de proyectos con productores. Taller para extensionistas. (Tools for the Formulation of Projects with Producers. Workshop for Extensionism). (Unpublished technical document, Collaboration Agreement INCA Rural—COLPOS). Texcoco, Estado de México. Available on https://www.scribd.com/document
10. Perico RE, Ribero MP (2002) Nueva Ruralidad. Visión de Territorio en América Latina y el Caribe. (New Rurality. A Vision of Territory in Latin America and the Caribbean). Cargraphics, S.A. IICA. Pp. 17–20. San José, Costa Rica
11. Sagarpa-IICA: Cuaderno técnico del marco de la política del nuevo extensionismo (Technical notebook of the new extensionism policy framework). San José, Costa Rica (2017)

Chapter 8
Navigating Sustainability Challenges in Africa: The Ogun State Racetrack, Motor Sports and Autopark Project

Teri Vivienne Okoro

Abstract Integrating sustainability into major or even smaller projects and the business operations of organizations is increasingly de rigueur. However, delivering sustainability in process, product and benefits raises differing challenges dependent on context, extent of alignment with pre-existing strategic objectives and definition. Multiple frameworks to review sustainability within projects also exist. This study utilizes a hybrid framework to review the delivery of sustainability in practice within the early stages of a motorsports tourism project in Africa and concludes with some recommendations for project management practice and further research.

Keywords Africa · Social responsibility · Environmental impact · Social impact · Project success · Sustainable project management · Tourism

8.1 Introduction

Sustainability gained greater visibility within project management just over a decade ago and about 5 years later, Silvius and Tharp [23] assessed their relationship as "*…still picking up momentum*". Today, it is acknowledged as a relevant component in assessing project success as IPMA [11], greater numbers of studies including dedicated journal issues exist and it is fully embedded into competence frameworks as APM [3].

Businesses, meanwhile, before and during this period have been delivering new projects and undertaking ongoing operations to deliver income and profits, that impact the environment and society to varying extents. Sustainability benefits where achieved are not always by design or explicitly through sustainability principles as indicated by Keeys and Huemann [14]. At the macro level, reflecting our collective responsibility, the 2015 UN Sustainable Development Goals (SDGs) to which all countries are signed up provide a framework against which development can be

T. V. Okoro (✉)
TOCA, Alpha House, Gatward Close, London N21 1AS, UK
e-mail: terilondon1@gmail.com

R. Cuevas et al. (eds.), *Research on Project, Programme and Portfolio Management*, Lecture Notes in Management and Industrial Engineering,
https://doi.org/10.1007/978-3-030-60139-3_8

planned and progress mapped over time; building on the earlier Millennium Development Goals (MDGs). Both SDGs and MDGs recognise pre-existing problems in a wide range of areas impacting people and their environment, also the slow progress in tackling and eradicating them.

The concept of sustainability (including within project management) has evolved and today there is greater consistency in global understanding reflected in the three pillars of sustainability—economic development, social development and environmental protection identified in the 2005 World Summit on Social Development. However, it is pertinent to be explicit in the use of the term within this paper.

Silvius and Schipper [22] in their literature review and impact analysis identified multiple meanings associated with sustainability within project management beyond the three key elements of economic, social and environmental. These include the time dimension which is also one traditional meaning attributed to the term, stakeholders—their influence (global or local) and extent of engagement or participation, also the ethical element as values and transparency, use of resources—as minimizing waste including of resources in production or through utilizing income not capital and finally risk reduction as safeguarding is more efficient than ameliorating damage. This paper recognizes the overarching concern with time, cost and scope impacting quality, acknowledges the three classic dimensions of sustainability and will additionally focus on the time dimension, stakeholders and refers also to the ethical dimension.

The status quo situation in Africa at its most pessimistic has been described by Kanayo and Patrick [13] as *"...the problem of social and natural capital deterioration in Sub-Sahara Africa is quite alarming without sufficient response mechanisms"*. On a generic level, challenges of embedding sustainability in Africa are related to elements as extreme poverty, rapid population growth rate, rapid urbanization, deforestation, environmental impact of extractive industries, rate of economic growth, rural development, climatic variability and natural environmental hazard in [1, 13, 26]. However, in relation to major projects as in Okoro [18] and specific entrepreneurial ventures as Fairer-Wheels [8] the situation is often more nuanced and context specific, as noted in van Marrewijk [27] and Silvius and Schipper [22] understanding is essential.

Majority of sustainability studies in project management have focused on construction (including extractive industries), IT and sustainable development with most within economically developed countries as evident in Silvius and Schipper [22]. There is a need to expand this narrow sector and locational focus. Furthermore, most learning from projects usually occur post-completion or, at the completion of a major phase. In-depth project review at the concept and early planning stages provide insight for external audiences and the project team, benefitting also future stages or phases of the project.

A key concern in this study is: In the context of challenges, how can the delivery of sustainability on projects be enhanced? With various sustainability frameworks to assess performance emerging within project management studies and more generic business studies, selection of an appropriate one is key for analysis. In the first part of this study, some frameworks will be reviewed prior to identifying the hybrid framework adopted for the analysis within this study. The methodology, case study context and findings are then discussed. In this context of emergence, learnings for future practice and studies are included in the final recommendations section.

8.2 Sustainability in Business and Project Management

Business has traditionally focused on returns to shareholders. This model has increasingly been called out since the 1990s by consumers, as well as academic, governments and environmental observers. Today, the focus is on the triple bottom line of profits, people and the environment. Referred to as Triple-P (people, profit and planet) or TBL in papers, there is a well-established body of work in this area as Elkington [6, 7]. Various terminology as corporate social responsibility (CSR) and sustainability is used without consistency but always with reference to the three classic elements as in Arowoshegbe, Emmanuel and Gina [4]. The extent to which sustainability and CSR are aligned to strategy also varies.

Its adoption also ranges from the mature where it is transparently tracked and publicly reported, to those with written policies not acted upon or ignored and others with no policy. In their analysis of factors contributing to traction in its adoption, the IFC in Sustainability, IFC and Ethos [24] identifies three myths that have hindered the adoption of sustainability principles in the world of business. The first is that it had no place in mainstreamed hard-nosed business and was for soft-hearted visionaries, the second is that sustainability efforts belonged downstream in the production cycle and the third which they considered most damaging was that sustainability only really made sense for rich companies in developed nations and the private sector in the emerging markets. Great efforts are being made today through a range of mediums including legislation, case studies and rewarding companies that operate on sustainable principles. There is still however much work to be done.

Sustainability in project management still has a myriad of definitions and dimensions. Silvius and Schipper [22] helpfully assess these (Table 8.1). Increasingly these are being assessed and collated into a structured format related to the Economic, Social, Environmental and other dimensions. These have relevance for impacting the sustainability dimension within project management.

Sustainability impacts project management throughout its life cycle. There appears to be limited agreement on the extent to which it impacts various activities. Half or more of the literature reviewed in Silvius and Schipper [22] agree on stakeholder engagement, project specification/deliverables, business case/deliverables, procurement, risk identification and project reporting. Different roles within the project team also perceive sustainability differently as noted in Silvius and Goedknegt [10]. While some as French [9] have charted a roadmap of sustainability within the project and programme management, the link between delivering sustainability and project management methodology is still not fully defined.

Another dimension requiring consideration is the context of the project. Just as informed consumers and public reputation can inform the extent of adoption of

Table 8.1 Dimensions of sustainability in project management literature (Adapted from Silvius and Schipper [22])

Dimension	Element	Comment
Economic		These also include balancing the social, economic and environmental
Social		
Environmental		
Other	Short or long term Local or Global orientation/drivers Ethical or values Transparency and accountability Stakeholder Participation Risk reduction Eliminating waste Consuming income not capital	

sustainability within the project, likewise the project team including its sponsors can contribute to its adoption, the direction of the project and its success. The presence or lack of patronage can assist unsustainable projects or derail beneficial ones. Within a messy or VUCA context, altered responses and adaptability can be anticipated as noted in APM [2].

It is recognized that challenges exist in implementing sustainability within projects in emerging countries and Africa, particularly, beyond the previously noted view among some of its inapplicable to their context. This includes lack of basic baseline data and ongoing measurement where change occurs. Without this, it is almost impossible to track progress on sustainability within projects as indicated in AUC, ECA, AfDB and UNDP [1]. Obtaining and collating data in these instances often from scratch is both time-consuming and more expensive. Shortage of local expertise and the need to hire foreign consultants adds further to the cost. Stakeholder engagement can often be patchy linked to identification and communication limitations as suggested in Okoro [17, 18]. The context of project delivery in Africa is a relevant additional risk or challenge to be navigated.

8.3 Framework for Assessing Sustainability in Projects

With sustainability in project management still evolving and varying frameworks in existence for assessing sustainability in projects, this section reviews a few including one originating from project management practice from Silvius and Schipper [22], another from project management with a strengthened business focus from Keeys and Huemann [14] and two from a purely business perspective from (Sustainability et al. [24] and PWC [20].

Silvius et al. [22] posit three shifts within sustainable project management occuring alongside the dimensions of sustainability (Table 8.1) that are present within projects—scope shift, paradigm shift and mind shift. These though focussed on the perspective of the project manager, are equally relevant for the rest of the project team.

In adopting a benefit and business value perspective, Keeys and Huemann [14] recognize that "*projects are challenged to meet individual organizations business objectives and value concerns*". They explore a framework within which stakeholders co-create to shape the sustainability benefits and value of projects particularly relevant for the long term as projects are transient and often have a shorter temporal focus and concern with more limited outputs and deliverables.

IFC within their Sustainability et al. [24] examination of businesses in emerging economies have developed a Business Case framework for the review of sustainability within projects and operations that examines the extent of alignment between the two. Utilizing a business case matrix enables a graphic heat map illustrating at a glance where a viable business case may exist. Their sustainability dimensions vary somewhat from those identified in Silvius and Schipper [22]. In their comparison of matrices of businesses in developed and emerging markets they identified that the latter, "*…focus more on short-term cost savings and revenue gains while reputational issues are more significant in developed countries.*" A visual matrix is useful for conveying information swiftly.

Additionally, various consultancies as PWC [20] have developed decision-making tools for businesses to evaluate the impact of business options on sustainability elements to assist informed decision-making. These tools are particularly effective where full datasets exist. In contexts with limited datasets the accuracy of the tool is questionable. Nevertheless, the principle of considering sustainability impact of options and in all decision-making and using appropriate tools is good practice to be built into processes. Their use post-decision-making can provide the benefit of learnings for the future.

The hybrid framework adopted for analysing sustainability in this case study adopts the sustainability dimensions identified by Silvius and Schipper [22], incorporates the co-created benefits and value suggested by Keeys and Huemann [14] but also includes grading of low, medium and high that can easily be converted into an at a glance heat map for reporting. Other elements as the extent to which the sustainability mindset is developed and its consideration holistically in approach and decision-making are useful in determining the sustainability maturity of the project and team. Granularity is essential for any framework to ensure it is context-relevant.

8.4 Methodology

The fundamental purpose of this study is to examine sustainability and its challenges in an auto tourism project and also establish ways to enhance sustainability on future phases and projects. This study adopts a primarily qualitative research technique, exploring the challenges of sustainability through interviews relating to a singular case study. The project type, sector and commercial nature limits the number of available study projects.

The interview questions covered the project context generally, together with a more focussed examination of sustainability within the project. Two semi-structured in-depth interviews occurred enabling the collection of substantial primary data. The data was transcribed and coded. The dimension of sustainability identified within the adopted analytical framework informed the coding as economic, and social.

Secondary data desk research of the context was also undertaken, to enhance understanding and also permit triangulation where possible.

Case studies can contribute to a better understanding of cause and effect relationships while permitting myriad of factors and their interaction to be reviewed. Their long recognized standing as an established research method Yin [29] has seen them favoured also for studying sustainability within project, including those with construction elements.

8.5 Case Study

8.5.1 Project Context

Considered an emerging economy, Nigeria with a population of 191 million is the most populous country in Africa. It accounts also for 47% of all people in West Africa and has an extremely high youth population (its national median age is 18.4 years). Its GDP per capita of 2,412 USD is tenth highest in Africa but low compared to 6,160 USD in South Africa, 16,433 USD in the Seychelles—the highest ranking African country and 59,532 USD in the USA as recorded by the World Bank [29]. The middle-class population is on the rise.

Nigeria's economy and foreign earnings have been dominated by oil production which represents 90% of its foreign income though only 9% of its GDP. Recent drops in global oil prices have impacted its income adversely, slowed down its economy, dropped its currency value and contributed to double-digit inflation. Unemployment is currently 23% Trading Economics [25]. There are growing efforts to diversify the economy including through tourism and to attract inward investment. Services account for half of its economy. Government capital expenditure impacts economic growth positively as highlighted in Jerilov and Musa [12] with election periods bringing uncertainty and an element of paralysis to enterprise and projects.

Tourism in Nigeria is driven primarily by the domestic industry which in 2017 accounted for 97% of spend and 5% of the GDP as stated in Deloitte [5]. In contrast, South Africa's tourism accounts for 9% and in the UK, it is just under 10%, therefore great potential to expand this sector locally and internationally exists in Nigeria, as also indicated in the PWC [21] leisure and hospitality outlook report. No formal auto sports leisure sector currently exists in Nigeria.

Motoring in the country is primarily for transportation, car ownership—the motorization rate (car ownership per thousand) is not accurately established but low at between 20 and possibly up to 60. Driving can be dangerous with a high incidence of mortality on the inter-city roads linked to poor driving skills, poorly maintained vehicles and in some instances bad road conditions requiring cars to use lanes designated for oncoming vehicles.

The auto industry in Nigeria is dominated by imports though it has a chequered history of local assembly production since the 1970s. There are no significant threats from local production in any neighbouring countries. High import tariffs have restricted imports during the recent recession. Imports have traditional included pre-used cars from Europe. Local production from the active domestic assembly plants–clustered in the capital Lagos State, with one plant in Ogun State and another in the north of the country–is currently low. There is a new NAIDP [15] initiative "*towards growing an automotive sector that serves the domestic market*".

Ogun State one of the 36 states in Nigeria, is located close to Lagos the country's largest city and economic hub, less than an hour's drive away. It is progressive, also the country's most industrialized state (e.g. cement, plastics) has good transport links and a good record on energy outages. It benefits from inward investment and is keen to diversify its industrial and largely agriculture income base. Reflecting its open for business ethos, the government has a One Stop Shop for new investors to simplify business set-up, facilitating permits, access to premises and land, etc. Fourth in state rankings for ease of doing business according to the World Bank [28], it aims to attract meaningful local, foreign and diaspora investment and become …"*an investment destination of choice*" [16].

8.5.2 Project Objectives, Strategy and Team

Against this national and local context, the Ogun Motorsports project is emerging. With a multi-layered strategy, the project has a long-term legacy vision. In addition to providing a regional racetrack and associated leisure destination, it also aims to be a centre of excellence for the auto industry both in education and manufacture [19].

Leisure—the flagship international standard racetrack aims to provide entertainment, expose and encourage youth to join the sport and educate them about fast but safe driving. Other auto facilities are motor cross, off-road rally and quad bike facilities. Ancillary leisure includes go-karting and polo. Retail and two commercial hotels (one with lodges set in the natural landscape around the enlarged lake at the quieter rear end) and some condos will service the park.

Training—the other Phase 1 focus is on raising the driving skills of commercial (including haulage, logistic, lorry) and other drivers is an important element to improve the safety standards in the country. This has garnered much interest from industries as the oil, gas, agricultural and other commodities. Training of the mechanics and motor engineers of the future is also planned. The training ground also provides for testing. A medical centre is to be provided.

Industry—is the final phase of the project, auto companies co-located alongside motorsport. Assembly plants, spare parts manufacture and sales also new car sales are planned. These have to be fit for and aligned to the future of motoring.

The various components of this ambitious project contribute to its vision to be the MotorCity of Nigeria, a centre of excellence and creating a lasting legacy. The sustainability review will focus primarily on the racetrack and leisure element as the auto park industry is a subsequent phase.

The project is a public-private venture, with the state forming a joint venture with the private sector consortium consisting of an international motorsports expert, a local motoring sector expert, an international marketing and project management company with local links and multiples investors most foreign but some local.

The Ogun State Governor is the sponsor and a key supporter of the project, there is a core project team with various consultants and suppliers also involved. Team members were selected for the skills and competences they bring to the project. The Project Director from the marketing and project management company assisted by a small project team, manages resources/budget/risk and was involved with preparing the business plan and feasibility study. In this instance, the Project Director is not subordinate to the sponsor and retains a greater level of agency than usual.

8.5.3 Research Findings—Sustainability in the Project

Initial ESIA—the Environmental and Social Impact Assessment and Action

Sustainability was a consideration on this project and the existence of challenges was apparent quite early. The initial ESIA screening was commissioned with a specific requirement to utilize local consultants where possible and additionally to allow for the transfer of skills to local consultants. The ESIA was required to meet international standards in order to meet the requirements for international investments. This highlights the first of several challenges the lack of local skills. The appointment of foreign consultants came with associated high costs for a project that was not fully funded at the early stage. The cost of travel, hotel accommodation and payment when the local currency value had dropped was exorbitant compared to similar projects in developed countries.

A further challenge was the limited availability of base data, reflecting an issue previously highlighted by Okoro [18]; AUC, ECA, AfDB and UNDP [1]. This increased the scope of the work to be undertaken and also extended the timeframe. With a high expenditure at an early stage of the project, it is unsurprising that the report has not currently been widely circulated beyond the key individual and the relevant authorities due to its commercial value.

The initial ESIA screening has contributed to skills transfer benefitting local consultants in the short term, as well as the future.

The social impact of the project is mixed though mainly positive, however, the five existing small settlements on the identified site will have to be considered and sensitively relocated, however, the inhabitants will benefit from a more secure source of funding from employment linked to the project.

The environmental assessment relating to flora and fauna has not highlighted any significant negative impact, which in no small way reflects considerations during the original site selection and the proposed design of new structures. The 500 hectare site is underdeveloped agricultural land, with dense vegetation occuring in some areas.

The site has elevated ranges which are factored into the outline design. Their potential as a natural viewing point for the racetrack has already been identified. Incorporating them into the overall design also eliminates the need for levelling works to them that will impact the pre-existing landscape and add to the project costs. Limiting earthworks and infrastructure costs has been a key design strategy of the project. A new slip road is, however, proposed for use during peak event period to minimize the risk of congestion.

The high local temperature has been taken into account and sheltered areas are being retained as far as possible as sheltered spectator areas. Existing water attenuation areas will be factored into the overall natural water drainage and management system for the surrounding area including the feature lake. This recognizes the heavy rainfall that occurs locally.

A pre-existing environmental element highlighted is the plastic pollution of waterways and generally. A sustainable solution to deal with this has been identified and is already being implemented in advance of the main project. The first element of the project is the recycling of the plastic, linked to income generation for the local population. The second is utilizing the recycled material in the fabrication of the racetrack providing the dual benefit of the lowered cost but also simultaneously reducing the travel distance of materials and carbon utilized.

Currently, off-grid, sustainable solar energy will form part of the energy provision.

The preliminary engagement has commenced with relevant stakeholders and this will be ongoing. Coverage of the project in the media has raised awareness locally.

Elections occurred in early 2019 and for a considerable period before and sometime after, while the new administration is settling in and formulating their strategies, the project programme has been delayed. This delay has resulted in the loss of one investor impacting the scale of the first phase. Elements as the recycling project are however still ongoing.

8.6 Results Discussion

8.6.1 Analysis

Utilizing the framework of benefits and value creation alignment with the sustainability dimensions, the following can be deduced (Table 8.2). The Ogun State motorsports project is commercially driven aiming to provide benefits for its investors as well as for the local government and the people it represents. While sustainability is

Table 8.2 Dimensions of Sustainability and Benefit/Value rating in case study project

Dimension	Presence	Benefit/Value—Public	Benefit/Value—Private	Comment
Economic	✓	Medium/High	Medium/High	Key driver
Social	✓	Medium/High	Medium	CSR for road safety is high
Environmental	✓	Medium	Medium	Value-driven benefits
Long term	✓	High	Medium	
Global driver	✓	Medium	Medium	
Stakeholders	✓	Low/Medium	Low	
Accountability	✓	Low/Medium	Low/Medium	

not a key strategic factor, its alignment with delivering business value and benefits to public and private partners has assisted in its delivery at this early stage of the project.

There is variance though some correlation in the value of sustainability linked benefits for the different partners and over time. For the private partners, benefits and value are largely economic or financial with social benefits as contributing to road safety, jobs and skills creation also growing the leisure and tourism provision. For the local government, regional development is the most prominent benefit, with additional economic and social benefits from jobs creation, income generation and skills development. Environmental benefits are largely neutral for both and unlikely to be negative. Reduced infrastructure works, retaining existing features and sustainable drainage are some positive facets; all aligned with value creation for the private investors.

Collaboration with the local government and their investor friendly stance has reduced some of the challenges of the context, benefitted the project and will do so going forward. In detailed analysis, local stakeholders will need to be separated from the local government. Stakeholder engagement and accountability are scored lowly currently and are low areas of value for the private partners, though going forward this is likely to improve, as necessary for the resettlement of communities impacted by the project. Areas of non-alignment in benefits and value could test the project in the future if the framework for decision-making is not robust and if relationships are weaker. Project management processes are not fully explicit including separate monitoring of sustainability dimensions.

Overall, benefits are apparent for the medium and long terms and have been achieved in numerous sustainability dimensions, indicating the possibility of enhancement with more dedicated focus and integration within project processes as occurred in the ESIA screening procurement.

8.6.2 Conclusion and Further Research

This study set out to assess how in the context of challenges, sustainability can be enhanced within projects. The consideration of a suitable framework to undertake the assessment of sustainability within projects was also a concern. For businesses and projects, a holistic approach that identifies the business case and value of sustainability at the outset assists in determining alignments as the project proceeds. Recognition of the various dimensions and continual review of decision-making implications through a context appropriate framework enables informed decisions by all partners. Areas of low benefit and value alignment can be risky and require close monitoring, as achieving sustainability is a continuous balancing endeavour.

Three future research suggestions are made. Firstly, the need to consider simpler sustainability frameworks for use by project managers and others delivering projects. Secondly, that consideration be given to benefits alignment challenges within projects of shorter duration or with greater negative impact—future exploration of the separation between benefits creation and capture as explored in Keeys and Huemann [14] could be one direction. Thirdly, an expansion in case studies and development of associated tools that cater for increased unpredictability within project operating contexts that require adaptability.

8.7 Compliance with Ethical Standards

All elements of this non-interventional social study comply with ethical standards applicable to research involving Human Participants, in keeping with those laid down by in the 1964 Helsinki Declaration and later amendments or comparable standards. It was reviewed by the Ethics Committee of P3MAfrica—ref 19003.

Prior informed consent was obtained from individual participants included in the study before interviewing. No sensitive personal data was accessed. Anonymity of individual participant data is maintained. Additional informed consent for the project identifying information included in this chapter (and prior Conference presentation) was obtained and the accuracy of data herein on the project process and delivery was cross-checked for accuracy.

References

1. African Union Commission (AUC), the UN Economic Commission for Africa (ECA), the African Development Bank (AfDB), and the UN Development Programme (UNDP) (2017) Africa Sustainable Development Report: Tracking Progress on Agenda 2063 and the Sustainable Development Goals,' Addis Ababa. Economic Commission for Africa. https://https://www.undp.org/content/dam/rba/docs/Reports?AGENDA2063_WEB_FULL-EN.pdf. Accessed 10 Aug 2019

2. APM (2018) BoK7: Project-based working in a 'VUCA' world. https://www.apm.org.uk. Accessed 4 July 2019
3. APM (2019) APM Body of Knowledge. 7th Edn. (eds) Murray-Webster R, Dalcher D Association for Project Management, Princes Risborough
4. Arowoshegbe AO Emmanuel U and Gina A (2016) Sustainability and Triple Bottom Line: an overview of interrelated Concepts in Igbinedion University Journal of Accounting, vol 2. August 2016, pp 88–126. https://www.iuokada.edu.ng/journals/9F2F7F6B56B433D.pdf. Accessed 10 Aug 2019
5. Deloitte (2018) Navigating the African Automotive Sector: Ethiopia, Kenya and Nigeria https://www2.deloitte.com/content/dam/Deloitte/za/Documents/manufacturing/za_Africa-Auto-2016-Report-28-May-2018.pdf. Accessed 8 July 2019
6. Elkington J (1997) Cannibals with Forks—Triple Bottom Line of the 21st century business. New Society Publishers, Stoney Creek CT
7. Elkington J (2013) Enter the Triple Bottom Line. In: Henrques A and Richardson J (eds) The Triple Bottom Line. Routledge, London, pp 1–16
8. Fairer-Wessels F (2012) Fostering Sustainability through entrepreneurship in South Africa: Selected Case Studies. Int J Multidis Bus Sci 1(1):28–35
9. French B (2017) Saving the world one project at a time. Presentation at the Australian Institute of Project Management 2017 Conference. https://www.aipm2017.com.au. Accessed 12 Aug 2019
10. Goedknegt D and Silvius AJG (2012) The implementation of sustainability principles in project management. In: Proceedings of the 26th IPMA World Congress, Crete, pp 875–882
11. IPMA (2016) IPMA Project Excellence Baseline for Achieving Excellence in Projects and Programmes. Version 1.0
12. Jerilov G, Musa M (2016) The impact of government expenditure on economic growth in Nigeria. Sacha J Policy Strateg Stud 5(1):15–23
13. Kanayo O, Patrick U (2013) The challenges and implications of sustainable development in Africa: Policy options for Nigeria. J Econ Cooperat Dev 34(4):77–111
14. Keeys LA, Huemann M (2017) Project benefits co-creation: Shaping sustainable development benefits in International J Project Manag 35(6):1196–1212
15. NAIDP (2013) National Automotive Industry Development Plan of NADDAC—National Automotive Design and Development Council. http://www.naddc.gov.ng/programme/automotive-plan. Accessed 5 July 2019
16. Ogun State (2018) https://www.ogunstate.gov.ng/onestopshop/. Accessed 22 July 2019
17. Okoro T (2018) Joining the dance: Creating an inclusive profession. APM
18. Okoro T (2019) For the public good: Facilitating Sustainability and inclusion through Environmental and Social Impact Assessments. In: Report on the 2nd Edition of the Pan Africa PMC
19. OSM (2017) Concept Design and Feasibility Report. Published by Ogun Motorsports
20. PWC (2013) Measuring and managing total impact: a new language for business decisions. https://www.pwc.com/gx/en/sustainability/publications/total-impact-measurement-management/assets/pwc-timm-report.pdf. Accessed 8 Aug 2019
21. PWC (2018) Hotels Outlook 2018–2022: Positioning for future growth—South Africa, Nigeria Mauritius, Kenya and Tanzania. https://www.pwc.co.za/en/assets/pdf/hotels-outlook-18-2022.pdf
22. Silvius AJG and Schipper RPJ (2014) Sustainability in project management: a literature review and impact analysis in Social Business, 4(1):63–96. https://doi.org/10.1362/204440814X13948909253866 Westburn Publishers Ltd
23. Silvius AJG, Tharp J (eds) (2013) Sustainability integration for effective project management. IGI Global Publishing, Hershey PA
24. Sustainability, IFC and Ethos Institute: Developing Value (2002) The Business Case for sustainability in emerging markets. https://www.ifc.org/wps/wcm/connect/topics_ext_content/ifc_external_corporate_site/sustainability-at-ifc/publications/publications_report_developingvalue__wci__1319577294013. Accessed 1 July 2019

25. Trading Economics (2019) Nigeria—Economic Indicators. https://tradingeconomics.com/Nigeria/indicator. Accessed 4 Aug 2019
26. UNECA (2011) New and Emerging challenges in Africa Summary Report (2011). https://sustainabledevelopment.un.org/index.php?page=view&type=400&nr=502&menu=1515. Accessed 4 July 2019
27. van Marrewijk M (2003) Concepts and definitions in CSR and corporate sustainability between agency and communion. J Bus Ethics 22(2–3):95–105 https://doi.org/10.1023/a:1023331212247
28. World Bank (2018) Doing Business in Nigeria 2018. https://www.doingbusiness.org/content/dam/doingBusiness/media/Subnational-Reports/DB_in_Nigeria_2018_w-bookmarks.pdf. Accessed 4 July 2019
29. World Bank (2019) The World Bank in Nigeria. https://www.worldbank.org/en/country/nigeria/overview. Accessed 4 Aug 2019
30. Yin RK (2017) Case study research and applications design and methods. SAGE Publishing

Chapter 9
The Role of Projects in the Process of Transforming Automotive Industry

Reinhard Wagner

Abstract The automotive industry is one of the most important industries worldwide. Since its beginnings at the end of the nineteenth century, it has been undergoing a transformation that essentially goes through four phases. Projects play a specific role for the process of transformation. The chapter explains the role of projects in each of these four phases and also looks at future developments. Projects are increasingly at the core of organizing innovation, product development, manufacturing and service delivery. This development, also known as "projectification", has strong repercussions on the organization itself. Organizational structures, processes and above all mindset and culture adapt to the requirements of project work. People are more clearly in the focus than before. Because companies can only be successful if they can respond flexibly to the new challenges coming from outside, i.e. from the market, competition and society. But that is precisely what distinguishes the automotive industry, after all, that it has been very adaptable and thus sustainably successful up to now.

Keywords Project management · Projectification · Automotive industry · Transformation

9.1 Introduction

The automotive industry is one of the most important industries worldwide. Last year, approximately 70 million vehicles were newly registered, most of them in China, Northern America and Western Europe. In Germany, more than 7 million people are directly or indirectly employed in the industry [8]. A structural or economic crisis in this sector is fuelling fears of job cuts or loss of competitiveness. Triggered by justified concerns about climate change and the responsibility of the automotive industry to contribute to the solution, increases the pressure on all stakeholders to change products, services and working methods towards more sustainability.

R. Wagner (✉)
Tiba Managementberatung GmbH, Perchtinger Straße 10, 81379 Munich, Germany
e-mail: reinhard.wagner@tiba.de

R. Cuevas et al. (eds.), *Research on Project, Programme and Portfolio Management*, Lecture Notes in Management and Industrial Engineering,
https://doi.org/10.1007/978-3-030-60139-3_9

With the help of the organizations in the automotive industry, mankind has been able to fulfil its dream of mobility, achieve undreamt-of productivity advantages, generate employment and prosperity and economic growth that is certainly unparalleled. The sector has also made a significant contribution to innovation, new knowledge and technological advancements that have been and continue to be used in other sectors [5]. With more energy-efficient engines, stringent exhaust gas treatment and lightweight construction, major steps have also been taken in recent years towards more sustainable transport. However, the pace of these changes today is far from sufficient to cope with the changed perception of cars in the eyes of society, strict legislation and the undesirable side effects of increasing mobility (e.g. traffic jams) as the population grows.

Since its early beginnings at the end of the nineteenth century, the industry has been subject to constant change. The invention of Carl Benz to motorize carriages formerly drawn by horses triggered a worldwide boom and can be seen as the starting point of the automotive industry. Incidentally, experiments with electric motors were carried out early on, but the combustion engine then prevailed for reasons of weight. This phase can be described as the "pioneer phase" of the industry, as the period at the end of the nineteenth century was strongly influenced by the inventions of pioneers worldwide. The first attempts at assembly line production were then made in the USA at the beginning of the twentieth century and perfected by Henry Ford. This triggered a huge boost in the industry, as large quantities could now be produced for a wider section of the population. As a result, attention was focused on rationalization, process optimization and the efficiency of operations. This phase can therefore be confidently described as the "industrialization phase". This phase brought tremendous efficiency gains, but these were bought by negative side effects such as unfavourable working conditions or unattractive vehicle models. With the bestseller "The Machine that changed the world" [21], Toyota suddenly became the focus of attention in the 1990s, a car manufacturer from the Far East that promised new growth with new, very flexible concepts. Until then, the main concern had been to optimize production, but now the customers' wishes for more individual vehicles became more and more the focus of attention. Work processes from the initial idea to delivery were organized in project form. In a ground-breaking article, Christophe Milder uses in 1995 for the first time the term "projectification" [14] to describe this phase and its developments. He highlights the effects on working methods, organizational forms and cultural changes at the French car manufacturer Renault.

The financial and real estate crisis in 2008 hit the automotive industry severely. Even though the industry was able to recover economically rather quickly, nothing has been the same since then. Global distribution battles, serious changes in societal values and trends, technological change and new, disruptive business models in the field of mobility have been shaking up the industry ever since. Suddenly, IT companies like Google are moving into the industry, start-ups such as Uber and Tesla are driving traditional companies like Volkswagen, Ford or Renault ahead, and in many places the car is even given a negative image and banned from the cities. That is why the current time could be labelled as the "phase of disruption".

In the following, the role of projects in the course of the development of automotive industry will be examined on the basis of the above-mentioned four time periods. Projects and project management are not only considered in a narrow, methodical focus, but also in an entrepreneurial sense, as a means of achieving organizational goals and strategies.

9.2 The Winds of Transformation and the Importance of Projects

In order to understand the role of projects in a given environment, it is necessary to look at the situation, the contextual factors and the various influences at that particular time. In addition to projects as a temporary form of organization, special attention should be paid to processes, technologies, business models, mindsets and culture. It is not the intention of this article to show every development in detail, but to give an overview and to provide a classification.

9.2.1 Pioneers, Ground-Breaking Products and Exclusive Customers

The success story of the automotive industry began with outstanding engineers. In Germany, Carl Benz, August Horch and Gottlieb Daimler are representatives of the pioneering achievements that were to shape a major industry. However, in the beginning they had to cope with many setbacks and the (economic) successes did not materialize for a long time. Engines were manufactured as a prototype in workshops, built on mobile bases and tested in a terrain that was supposed to be for horses and not for automobiles.

Since the vehicles were rather custom-made and not yet available in large quantities, the definition of "project" applies pretty much to what the pioneers did. Even though they certainly did not call it like that. The pioneers gathered around them a small number of experts who had to take on special tasks. Cooperation with other workshops or companies was sought only very hesitantly, as everyone was afraid of losing know-how and therefore wanted to keep the group of insiders relatively small. Since work at this stage was naturally capital intensive, the pioneer needed a sponsor who could provide money for the purchase of materials, tools and machines, and in this context the first stock corporations were established.

At that time only few people could afford cars. It was therefore reserved for the upper social class to gain experience with the first cars. In return they were willing to invest money in the development of more powerful engines and "car races" became popular, which promised new impulses and money for the development of engines. These developments began in Europe, but the spark quickly jumped over to Northern

America, where the new technology was intensively studied and own developments were initiated.

The role of projects in the pioneering phase of the automotive industry was therefore to combine new technologies in a workshop with a few employees to create a vehicle and satisfy an exclusive clientele. We don't know much about the methodology at that time, project management in today's sense was not yet available, but "managing" projects still had to be goal-oriented, pragmatic and based on experimentation, as little experience was available [10].

9.2.2 Taylorism, the Quest for Efficiency and Controlled Product Diversity

Assembly line production was already known in Europe and the USA in the production of food before it was applied in the USA for the production of vehicles, first at Oldsmobile and then by Henry Ford. The US American Frederick Winslow Taylor studied the work processes in factories specializing in assembly line production and founded the principle of "process control" of work processes. A clear distinction was made between "brainwork" of specialized planners and "manual work" of the ordinary workers. The process control is prescribed in detail by management based on work studies and work preparation. The term "Scientific Management" was later coined for this.

The procedures of scientific management according to Taylor are based on very detailed studies of work processes. The planning department followed a "one best way" approach and the workforce was instructed to follow it exactly. This has led to a "dehumanization" in the processes of the factory, people were recipients of orders and could not unfold themselves. The paradigm of planning was oriented towards efficiency in order to achieve the greatest possible output with as little material, capital and people as possible. As the market demand for automobiles increased, it was important to achieve larger quantities. However, diversity fell by the wayside. Henry Ford has been quoted as saying that customers can have a Ford in any colour, as long as it is black [3].

Even in this phase of the automotive industry, little attention was paid to "projects". The focus was instead on the ideal workflow organization or efficient processes. In contrast to the previous phase, the processes were split into planning ("white collar") and execution ("blue collar") work. The planning was based on mathematical optimization approaches and graphical representations. Henry Gantt, a student of Taylor's, also became famous in project management for his Gantt charts. Although this form of presentation was invented much earlier, it shows the chronological sequence of dependent steps in an overall task decomposed into individual components.

The optimization of production processes can of course be seen as projects that lead to a continuous stream of improvements from analysis to individual optimization measures. The avoidance of errors, approaches to quality improvement and methods of "Operations Research" had a strong influence on the development of today's project management. This was revised mainly in the USA in the context of military projects; network planning techniques, critical path method and other analytical approaches mark the beginning of project management, which is still strongly influenced by this time [15].

9.2.3 *Projectification, Mass Customization and Global Supply Chains*

The oil price shock in the 1970s led to a rethinking in the economy and also in the automotive industry. Customers wanted more individual products and manufacturers responded by broadening their portfolio of products. Naturally, a broader product portfolio required a boost in development activities and greater flexibility in production, which had previously been geared more towards large quantities. Suddenly, an unknown Japanese manufacturer moved to the forefront, which, with its Toyota Production System (TPS) and a flexible response to demand ("mass customisation"), was not only more profitable but also able to react much more flexibly to the dynamic developments on the global market.

The starting point for Toyota is the expectations of its customers. Translating these into requirements for the development and manufacture of vehicles and still practicing the greatest possible efficiency and cost savings became the famous TPS [16], which is still replicated in many companies today. Toyota uses the term "project" rather sparingly and mainly in connection with development activities that are a "series of problem-solving cycles" [4]. Nevertheless, the term "project" has gained new attention and a new meaning in the context of balancing the many development activities ("portfolio of projects").

Toyota's new management approaches have also been closely monitored and partially implemented by western car manufacturers. Christophe Midler analysed the developments at Renault over a longer period of time and in a remarkable contribution he coined the new term "Projectification" to describe the development. On the one hand, this refers to the increasing number and importance of projects in some companies and, on the other hand, to the transformation these companies are undergoing due to the greater importance of project work. Midler desc "The firm affected a transition from the classical functional organization in the 1960s to project coordination in the 1970s, and since 1989 to autonomous and powerful project teams. Search advanced project management has profound and destabilizing effects on the other permanent logics of the firm (task definitions, hierarchic regulations, carrier management, functions and supplier relations). The "projectification" process is still under way, to adapt these permanent processes to the new context" [14].

Midler states that not only is the role of projects increasing, but also the role of project management is becoming more and more important, moving from a rather coordinating function to the centre of product development. As a result, the organizational allocation is also changing and formerly subordinate project managers and teams ("temporary organization") suddenly become more powerful and thus come into competition with the powerful departments ("permanent organization"). As result, matrix organizations, project management offices (PMOs) and autonomous organizations for innovative vehicle models are emerging. Studies by the German Project Management Association show that the share of project work in total working time in the manufacturing industry has now reached almost 50% [7]. Project work dominates value creation, even far beyond the boundaries of a single company across the entire supply chain, i.e. internationally or even globally. Companies are developing into project-oriented enterprises that place projects at the centre of economic activity. Strategies, structures, processes and cultures change within the framework of project orientation [13]. Competence in project management is central, not only for those directly involved in project work, but also for executives and suppliers of the projects, right up to top management.

9.2.4 Disruption, Co-creative Networks and Innovative Mobility Solutions

The years from 2008 to 2010 mark the end of an ongoing upswing in the automotive industry. Instead, manufacturers and suppliers find themselves in a state of intense change, which is primarily based on global competition between the parties involved, disruptive technologies and business models, and socially altered value concepts, to name just a few of the factors. Increasing climate change with its consequences for the environment and society is calling into question the business of the entire automotive industry.

The "Diesel scandal" brings the industry into disrepute and stimulates a rethink in politics and society, which increasingly demand alternative drives or mobility solutions. The comet-like rise of start-ups such as Uber and Tesla also coincides with this period. At times, both achieve a higher valuation on the stock markets than all German car manufacturers combined. Uber does not have a single vehicle but uses digital processes to manage its collection service worldwide. Tesla focuses early on full-electric vehicles and the batteries they need. Initially smiled at, other manufacturers such as Volkswagen quickly follow suit and change their model policy away from the combustion engine and towards fully electric vehicles.

Autonomous driving solutions, alternative concepts for urban mobility and other innovative ideas require companies to rethink the way they work themselves as well as the way they cooperate with other companies. Hierarchical structures are too slow for a dynamic and increasingly complex environment, decisions must be taken locally, by professionals with access to the markets. Development processes require agile, self-organized ways of working by people with outstanding professional, social and personal skills. Co-creative networks with complementary partners are the "new normal" for project work. Project managers are network managers and moderate the numerous participants towards the common goal.

Increasingly, the product of the industry, i.e. the vehicle, is getting out of focus. Services, such as car sharing solutions or carpooling, are bringing new opportunities for value creation beyond the manufacturing of vehicles. The use of energy from the battery of vehicles helps to cover the electricity demand in cities in times of low energy production. The connectivity between vehicles, mobile devices and the environment creates additional value adding opportunities. Previously separate technologies and sectors are mutually beneficial and help to create more sustainable solutions for the environment and society.

Digitalization, Industry 4.0 and many other developments require a transformation of the automotive enterprise [2]. Projects and programmes are suitable forms for these changes, not only for technological but also for organizational transformations. Companies are becoming agile businesses, scaling agile working principles from individual projects up to the level of strategic business development and turning the entire corporate management upside down [18]. In the next section, the new demands on those involved will be dealt with in more detail.

9.3 Emerging Requirements for People, Organizations and Society

The changes in the automotive industry described above present new challenges for people, organizations and, beyond that, society. On the one hand, the individual needs a new mindset in order to better deal with the developments, e.g. openness and flexibility in dealing with changes [20]. On the other hand, concrete competences are needed for project work, which not only include the frequently taught methodology, but above all social and context-related competences, as described in detail in the IPMA Individual Competence Baseline (IPMA ICB) in its current version [11]. Qualification and certification in project management, agile or classic, plays an increasingly important role and is fostered by corresponding efforts on part of the companies.

But the transformation is also hitting companies with full force. Among other things, conventional ways of working, organizational structures and even familiar management principles are being called into question. If one starts to put people more clearly at the center of value creation in the company (as in the pioneering phase described above), they can regain motivation and respond in a competent and flexible manner to possible changes in the context of the project.

The company enables self-organization in the form of projects in order to co-creatively develop new and market-oriented solutions together with customers, partners and internal stakeholders [12]. At Tiba Managementberatung GmbH, together with companies from the automotive industry, new concepts for project work have been developed that offer answers to the challenges of our time. Projects in the automotive industry must meet the requirements of two worlds: On the one hand, work with a repetitive character, which requires classic project planning and control principles in near-production areas. On the other hand, the innovation work, which requires rather agile approaches. Both are not conceivable in isolation from each other but must be harmonized as optimally as possible. Digitalization and artificial intelligence methods are used for this purpose. In this respect, we also call this next stage in the evolution of project management at Tiba "Project Management 4.0" [1].

A prerequisite for agile work is certainly also a change in the set-up of the company. Voluntary action of employees requires clear space to manoeuvre as well as a protected environment, which allows new ideas to be experimented and tried out. In order to include as many perspectives as possible in finding solutions, a complementary team ("real team") should be formed, which works iteratively on the solution concept until the so-called "wow effect" occurs ("iterate to wow"). If possible, this is done with the involvement of the customer, so that immediate feedback can be incorporated into the solution and a real customer benefit is achieved ("customer benefit"). The coordination with the strategic orientation of the company as well as with all parties involved in the project is of course important and takes place via regular meetings, intensive communication and a close exchange with higher levels of the company ("mutual tuning"). After all, the biggest change takes place in management. An authoritarian leadership-style and micro-management no longer have a place in this environment, as they demotivate the team and give only a limited understanding of how to find solutions. True leadership supports the team in its work, sets a broad framework and helps with necessary resources and advice ("supportive management"). The latter of the seven "Tiba Principles for Agile Work" [9] is certainly one that places the greatest demands on change, as it scratches at the previous self-image of managers.

Nonetheless, companies that want to be future proof must face the challenges and transform themselves in the above-mentioned sense. This is done based on a detailed analysis of the requirements of the market, technological developments and the expectations of the employees towards the company. The latter is particularly important because "Generation Z" makes high demands on companies regarding the working environment and can often choose between several job opportunities.

But society is also facing new challenges in the volatile environment of the automotive industry. Employees and companies need a reliable legal framework within which they can operate. New forms of temporary work in projects not only have positive aspects. Work and leisure time are increasingly blurring in global project teams. Projects are an interesting field of activity for motivated people, but project managers often overload themselves in this situation and run the risk of burnout [6]. New forms of work require the acquisition of new skills, e.g. in dealing with digital technologies or in co-creative project work. This is not only a task for companies but is also a responsibility of society.

9.4 Conclusions and Outlook

This article dealt with the serious changes that the automotive industry has undergone in the more than one hundred years of its existence. Projects and project management in different phases of development sometimes play a more or less important role. While projects were unconsciously the organizational form for the pioneers at the beginning, project work has been pushed into the background by the Taylorism and the quest for efficient processes in the first half of the twentieth century. Only after the oil price shock in the 1970s and the reorientation of the industry did project work experience a renaissance through "projectification". Today, the automotive industry is strongly projectified compared to other industries. Projects are not only the preferred form of developing new products and services, but are also used for changes in the organization and corporate culture. Project management is trendy. However, project management has also changed a lot. In addition to classic, plan-driven project management, the importance of agile approaches is growing significantly. This also demands changes in strategy, processes, structures and culture.

Change and transformation happens within the scope of projects. In this respect, the right mix of project and change management is of great importance. Managers should understand that projects are not only a way to create innovative products and services, but also to successfully implement their business objectives. According to a recent survey by the Project Management Institute (PMI), top managers consider organizational agility to be an essential prerequisite for future success, along with investing in the right technologies and ensuring relevant competences [17].

In future, projects will be used in all areas of an organization, in addition to the more technical areas, also in HR and finance, supply chain management and associated networks of partners. Developments in the area of "Smart Cities", for example, show that a large number of stakeholders are coming together in projects to enable or further improve mobility. On the one hand, the city government, urban planners and public administration are called upon to create the prerequisites for mobility in their cities and urban areas. Subsequently, a large number of companies, such as automobile manufacturers, suppliers and innovative service providers, work together with the users to bring appropriate mobility concepts to life [19]. In the future, therefore, there will be no limits to project work. It will permeate all areas of our lives and society.

References

1. Balser T, Projektmanagement 4.0—Wie verändert sich die Welt des Projekt-managements. https://www.slideshare.net/TibaGroup/projektmanagement-pm-40-de. Accessed 15 Feb 2020
2. Brecke J, Nazareth D, Niederberger D, Ramsauer H (2017) Transformation von Automobilunternehmen. Books on Demand, Norderstedt
3. Ford H (2018) My life and work. Reprint. CreateSpace Independent Publishing, Scotts Valley
4. Fujimoto T (1999) The evolution of a manufacturing system at Toyota. Oxford University Press, New York

5. Gottschalk B, Kalmbach R (eds) (2006) Mastering the automotive challenges. Süddeutscher Verlag, München
6. GPM: Burnout-Gefährdung bei Projektmanagerinnen und Projektmanagern. GPM, Nürnberg (2014)
7. GPM: Makroökonomische Vermessung der Projekttätigkeit in Deutschland. Deutsche Gesellschaft für Projektmanagement, Nürnberg (2015)
8. Hab G, Wagner R (2017) Projektmanagement in der Automobilindustrie, 5th edn. Springer-Gabler, Wiesbaden
9. Koschke A (2019) How does the „agility“ trend affect project management?. Whitepaper. Tiba, München
10. Kozak-Holland M (2011) The History of Project Management. Multi-Media Publications, Oshawa
11. IPMA: IPMA Individual Competence Baseline (IPMA ICB) Version 4.0. IPMA, Nijkerk (2015)
12. Lang M, Scherber S (Eds) (2019) Der Weg zum agilen Unternehmen. Strategien, Potenziale, Lösungen. Hanser, München
13. Lang M, Wagner R (eds) (2019) Der Weg zum projektorientierten Unternehmen. Hanser, München
14. Midler C (1995) Projectification of the firm. The renault case. Scandinavian Manag J 11(4):363–375 (1995)
15. Morris P (2013) Reconstructing project management. Wiley, Chichester
16. Ohno T (1988) Toyota production system. Beyond Large-Scale Production. Productivity Press, New York
17. PMI: Pulse of the Profession 2020. Ahead of the Curve: Forging a Future—Focused Culture. PMI, Newtown Square (2020)
18. Tuczek H (Ed) (2016) Landshut Leadership. Führung im Zeitalter der Digitalisierung. Volume 1. Shaker, Aachen
19. Wagner R (Ed) (2018) IPMA Insight No. 1—Realizing Smart Cities through Professional Project, Programme and Portfolio Management. IPMA, Nijkerk
20. Wald A, Wagner R, Schneider C, Geschwendtner M (Eds) (2015) Advanced project management (Vol. 4). Flexibility and Innovative Capacity. GPM, Nürnberg
21. Womack JP, Jones DT, Roos D (1990) The Machine that Changed the World. How Japan´s secret weapon in the global auto wars will revolutionize western industry. HarperCollins Publishers, New York (1990)

Chapter 10
Walking the Talk? Sustainability in New Product Development Projects in the Icelandic Seafood Industry

Inga Minelgaite, Bjarnveig Birta Bjarnadottir, and Kari Kristinsson

Abstract The purpose of this research was to investigate new product development projects in the Icelandic seafood industry regarding three sustainability dimensions, namely, environmental, economic and social. Iceland is ranked among the most sustainable countries in the world. Furthermore, the seafood industry in Iceland, with its start-ups, has been referred to as a role model in the use of fish waste to develop new products. Using survey methodology, we investigate how the sustainability dimensions are represented in each phase of new product development projects in the seafood sector. The research contextualizes sustainability in new product development projects in regard to a company's strategy. The main findings demonstrate that the different stages of product development do not differ in aspect to sustainability. The case is particularly interesting from the perspective of sustainability in new product development projects, providing insights that can be applicable for other industries.

Keywords Sustainability dimension · Project management · Product development · Seafood industry

10.1 Introduction

Sustainability has become a significant topic in modern societies. The sustainability concept refers to the avoidance of damage and depletion to natural resources in order to maintain an ecological balance, while making sure that the generations to come have an equal ability to meet their needs [34]. Sustainability is generally

I. Minelgaite (✉) · B. B. Bjarnadottir · K. Kristinsson
School of Business, University of Iceland, Sæmundargötu 2, 102 Reykjavík, Iceland
e-mail: inm@hi.is

B. B. Bjarnadottir
e-mail: bbb11@hi.is

K. Kristinsson
e-mail: karik@hi.is

R. Cuevas et al. (eds.), *Research on Project, Programme and Portfolio Management*, Lecture Notes in Management and Industrial Engineering,
https://doi.org/10.1007/978-3-030-60139-3_10

considered to have three dimensions—environmental, social and economic. Through growing awareness of sustainability issues, organizations are also realizing the major influences, both positive and negative, they have on societies. By changing their strategy and incorporating sustainable management techniques, organizations can have direct societal influence with the services and products they are providing [12], which has been partly reflected in the growth of the Corporate Social Responsibility Movement [1]. The important value of sustainability for consumers is therefore gaining vast weight in the product development process [27]. However, with the vast difference in assignments and responsibilities within the product development process, the commitment to sustainability from one phase of the process to the next can also vary considerably.

The Icelandic marine sector is particularly interesting from the perspective of sustainability in new product development projects, due to its size and environmental, societal and economic impact on Icelandic society. It is one of Iceland's largest industries with vast job creation from fishing to processing, with machineries, services and technological innovations, transporting to sales and marketing, as well as endless product innovation. Moreover, the Icelandic marine sector has an endless opportunity for product development.

First, we will introduce the literature review concerning both sustainable product development, the Icelandic Marine sector as well as the different stages of product development. Then we will demonstrate how the investigation was carried out by collecting data from personnel at each stage of the new product development process within the Icelandic seafood industry and other organizations within the Icelandic marine sector. Employees and managers were invited to participate in a survey questionnaire where they would state their place within the product development stages as well as their responsibilities and commitments to sustainability. The purpose of this research was to investigate new product development projects in the Icelandic seafood cluster regarding the three sustainability dimensions. We investigate how the sustainability dimensions are represented in each phase of new product development projects in the seafood sector, shedding light on the different sustainability commitments from one stage of product development to the other. Therefore, the main purpose of this study was to investigate how the commitment to sustainability and sustainable product development are different between the four levels of the new product development process, with the hypotheses that commitment to sustainability varies from one level to the other

10.2 Literature Review

10.2.1 Sustainable Product Development

The term and idea of product development is represented in different ways in the literature, where the idea of different developmental stages seem to vary. One stream of theory suggests that the product development process spans through marketing, design, management and other activities done between defining a market opportunity and starting production [5]. Others say that the process is a five-gate system including scoping, building a business case, developing, testing and finally launching a product [7]. Nevertheless, there seems to be a general agreement that product development includes a manufacturing process, and its supply, distribution and support systems which lead to a final product. By contrast, Veryzer [30] argued that product development constructed in the context of a high degree of technological and market uncertainty does not function via a formal process. In addition, the author argued that innovative product development processes were more exploratory, and less customer focused.

Necessary factors for product development are both initiative and creativity, which often lead to the creation of something new that has not existed before, at least not in the form being created in the product development process [5, 7]. The different responsibilities between the five phases of product development imply that each phase has objectives and activities that must differ. Numerous research studies have addressed the many different projects that must be completed within the development process [14, 19, 23, 30]. This indicates that the responsibilities and commitments to sustainability must also differ between the stages of product development. According to Cooper and Edgett [8], the development and deliverance of new products that are differentiated, solve major customer problems, and offer a compelling value proposition to the customer is a key factor in success and profitability. However, many of these ideas fail during the developmental stages. Often the faults are missed steps and activities, poor organizational design and leadership, with inadequate quality of execution [7]. Trimi and Berbegal-Mirabent [29] argued that even though it was once believed that products were only developed as the result of a technological breakthrough or a disruptive technology, that creating differentiated, high quality products begins with understanding customers' unmet needs and value orientation. That being said, they also argued that managers should be able to respond to technological change, realizing how to deliver new technology to customers, requiring businesses to constantly evaluate their value propositions to ensure their offer matches well with customers' demands.

Moreover, the development of products with low environmental impact is one of the key contemporary trends in environmental management, and therefore managers need to consider integrating environmentally friendly technologies and inputs throughout the life cycle of products [21]. New product development with properly integrated sustainability dimensions has reportedly been beneficial to organizations through greater efficiency in resource use [26], enhanced corporate image [6], increased sales and market share and greater know-how concerning the creation of new technologies [9]. Furthermore, the development of environmentally sustainable products can positively influence operational performance [16], innovation [15] and market performance [13, 25]. The economic benefits of sustainability often follow on from the environmentally sustainable benefits. That is to say, optimizing raw material and energy usage, reducing consumption of other resources and improving waste management often results in reduced production costs. Enhanced operational and market performance through sustainable product development can also result in greater organizational value creation [24]. Social sustainability is often referred to as the general well-being of employees instead of the influence organizational transactions could have on their local society [3]. There are, however, a lack of theoretical and empirical studies regarding social sustainability and organizational management [11].

The personal values of employees and managers, their attitudes and commitments to sustainability also have a significant weigh in how committed organizations are towards reaching their sustainability goals [22]. Katsikeas, Leonidou and Zeriti [18] also found that the effects of eco-friendly product development on product development effectiveness become weaker when the organizational environment was complex. Organizations tend to limit their product strategies in complex environments. Managers therefore need to design a structure for sustainability commitment across all areas of their organization. This structure should provide both available and qualified resources on each phase of the development processes to facilitate effective environmental management. That is because achieving sustainability will require not only attention to the technical details and scientific enhancement of systems but also the human element, which varies from project to project and from person to person [10]. Furthermore, Leonard-Barton [20] argued that core capabilities could possibly hinder innovation. Highly innovative organizations must challenge their current capabilities and methods in order to advance and develop.

10.2.2 Icelandic Marine Sector

One of the Iceland's largest industries is the marine sector, which has stimulated considerable job creation and innovative product creation. This includes activities ranging from fishing to processing, with machineries, services and technological innovations, transporting to sales and marketing, as well as endless product innovation. One fish can provide an array of products, including seafood, canned goods, dried fish, marine supplements, cosmetics and medical products, fish leather and omega-3 oils. Over the past few years, there has been a notable development within the Icelandic marine sector. Importantly, 2018 was the first year when exports of technological machinery and services outweighed exports of Icelandic cod, demonstrating the importance of the marine sector as a whole. The Iceland Ocean Cluster with its innovative incentives has played a vital part in economic growth in Iceland. Even though job creation in fish processing has decreased over the past few years, innovation through the marine sector has created many new jobs. Fish farming is becoming a large industry and start-ups have emerged focusing on everything from fish-skin wound treatment to beauty products made from collagen. As has already been demonstrated, there are vast opportunities from fish by-products.

Different manufacturers take varied approaches to implementing their environmental practices for sustainable product development depending on various driving factors such as customer awareness, legislation, economic benefits and competitive strategies [17]. However, due to influences from various stakeholders, including government, consumers, societies and the business partners, attention to sustainability needs to be paid. Even though there are commonalities in new product development projects in the context of sustainability, however, there are also contextual variations. For example, Kara, Ibbotson, and Kayis [17], in their large-scale research project, found out that geographical differences result in different legislation, competition and consumer pressure to utilize various decision tools.

There are a few design tools in the literature that are discussed in the context of sustainable product development. Probably the most common ones are LCA (life cycle assessment), DFE (design for environment) and ECQFD (environmentally conscious quality function deployment). We will address each of those in turn. An approach of material grouping for simplified product LCA has been presented by Sun, Rydh and Kaebernick [28]. Their approach is essentially a simplified method to evaluate the environmental impact of material selection for products. It is based on the analysis of the materials they use, which are then classified into groups according to the type of material and their environmental performance. Vinodh and Rathod [31] argued that integrating LCA methods can facilitate sustainable product development.

The basic idea behind the DfE methodology is to bring environmental expertise directly to the designers [4]. The introduction of DfE methodologies in manufacturing minimizes future weaknesses and inefficiencies by paying attention to the environmental aspects right from the start of the design stage, leading to a reduction in both materials used and the waste of products. DfE ensures minimized potential environmental impact throughout the life cycle of the product by allowing innovators not only to create just the product but have influence during its whole life cycle [33].

ECQFD centres around a design methodology for an environmentally friendly product design. By classifying the significant parts of improving environmental consciousness and identifying the impact of design changes with respect to potential environmental improvements, ECQFD can be used as s a tool or a method at the product development stages to ensure sustainability [32].

Design tools such as LCA, DFE and ECQFD approaches are likely to be utilized in the medium- and high-complexity product development by, e.g. OEM (Original equipment manufacturers), as suggested by Kara et al. [17]. However, in order to use these approaches, it is essential that the company or industry has an organizational know-how about application of these approaches. Furthermore, often these tools are used with a focus on the operational part of the project, that is how it will be applied in the operation (e.g. production line) once the product is developed, or in other words, once the project of product development is finished. Each product development is unique, has beginning and end and scope, and hence by definition it is a project. New product development projects are new every time and require a way to maintain sustainability as an underlying value, regardless of the project´s unique nature. To sum up, even though there are some tools that enhance sustainable thinking in new product development, in order to actually apply those tools, companies need to (1) be educated about those tools, (2) establish ways how those tools are used not just in operations, but also in one time endeavors—development projects, that are unique every time and (3) make use of those tools implicit in an organization´s mentality and processes. Applying knowledge from other fields or other geographical locations is not simple, as particularly large industries are face unique and different situations of sustainability in each geographical region or social context. This is why in this paper we focus on a particular industry in a specific context—Iceland, when investigating sustainability in new product development projects.

10.3 Methodology

We collected data from personnel at each stage of the new product development process within the Icelandic fisheries and other organizations within the Icelandic marine sector.

10.3.1 Procedures

Data for the main study was collected over a period of one week in April 2019, where the researcher visited 40 possible organizations which were identified from online research as being organizations within the marine sector associated with product development. Each individual was asked about his or her job title and daily projects before answering the questionnaire. None of the individuals acknowledged that they had any formal training in sustainability or product development. Moreover, based on the researcher's short interaction with these individuals, most of them either had degrees as engineers or technologists. Their responsibilities towards product development were often intertwined with other responsibilities of administration and operations, and often the development projects were their responsibility by pure chance. Only one organization, the Icelandic Ocean Cluster, was given advance notice of the arrival of the researcher. Within the Icelandic Ocean Cluster are about 70 different organizations, which is why they were contacted beforehand. Other organizations who participated received a short introduction minutes before answering the questionnaire. Due to the unique sample this approach was seen as the most feasible option, it had to be made sure that only individuals within the marine sector and involved in product development were participants.

10.3.2 Participants

The target populations for this study consisted of individuals who were related to product development within the Icelandic marine sector. The sampling frame was a list of many, if not all organizations, within the marine sector in Iceland, which were found through online research. Organizations were deemed unsuitable and excluded if they did not participate in product development. Purposive sampling was used to gather data which lead to the sample being specifically chosen based on their job title and day to day projects at work. Each individual necessarily had to have connections with at least one of the four stages of product development in the Iceland marine sector.

Initially, there was not a determined target sample size. The marine sector is one of the largest industrial sectors in Iceland, with various service and research firms coupled with an initiative through the Icelandic Ocean Cluster. There were about 110 organizations within the marine sector reported as actively participating in some kind of research or product development. Within these 110 organizations, we estimated that each organization would have about three individuals working in product development drawing that conclusion on research and development departments from some of the organizations. Of that, the sample reported here contains only about 40 organizations, which gives us a response rate of about 36%.

A questionnaire was delivered to these 40 different organizations and the researcher handed them out and received responses. There were 75 questionnaires handed out and 65 were answered entirely. The sample was dominated by male respondents, 74% and the majority of the respondents fell in the 45–60 age group. The educational level was relatively high and approximately 69% of the respondents had a higher educational level of either a bachelors-, masters- or a master craftsman's degree. The industry is general is dominated by males at 85% of job participation. However, over the industry of agriculture and fisheries only 7% of individuals have a degree of higher education. No measures were undertaken to match the responses to typical responses from individuals working in the product development of other industries.

10.3.3 Measurement Scales

Figure 10.1 explaining four different stages of product development was used to assess the different phases of product development that our respondents were responsible for. A simplified nominal scale of Cooper [7] product development process was used where respondents would choose between four different stages. The first stage being Needs assessment and research, the second being Conceptualization and development, the third item was Production, with the final item of Sales and marketing.

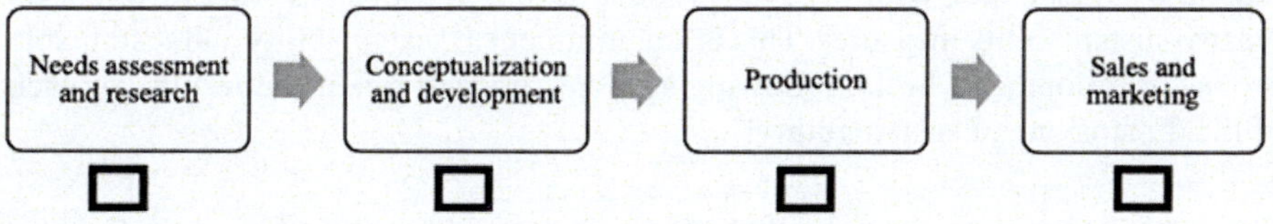

Fig. 10.1 A simplified version of Cooper's product development process

This first step of the questionnaire was done in order to understand the different responsibilities towards sustainability that each participating individual had.

A simplified version of Alcock's [2] measure of commitment to sustainability was used to assess the respondent's personal values concerning sustainability. The first five items of the measurement were five-point Likert scale statements. All the scale points were labelled ranging from 1 ("Strongly agree") to 5 ("Strongly disagree"), with the middle statement being neutral. Alcock's measure was particularly fitting since the author seeks to measure commitment, not attitude, by setting sustainability in the context of personal cost with time, effort and changes to lifestyle.

For measuring sustainable product development, a scale for sustainable consumption was adjusted to fit statements of product development. To distinguish from economic, environmental and social dimensions of sustainability an original scale from Balderjahn et al. [3] was used. The scale can be divided into three sections, one section reflecting the economic pillar of sustainability, the second section reflecting the environmental pillar and finally a section on the social pillar of sustainability. However, the scale did not provide a good measure for social sustainability. Each statement related to social sustainability emphasized employees and workers, not addressing the topic of gender equality or the possible social impact of the product. The original scale from Balderjahn et al. [3] had statements such as "Is it made from recycled materials?", "It is produced in a manner which minimizes the use of resources?", "Workers' human rights are adhered to?", and "It is a useful product?". These statements were adjusted to: "We use recycled materials.", "We minimize the use of resources.", "We obey laws of human rights.", and "We bear in mind that the products and services are useful." This 12-item, five-point Likert scale measured the respondents' professional attitudes towards sustainability. The scale points were labelled from 1 ("The statement corresponds very well with us") to 5 ("The statement corresponds very badly with us"), with 3 beings ("This statement corresponds neither well nor badly with us"). In addition to the above-mentioned items, the questionnaire also contained questions regarding the respondent's age, gender, living area and educational level. These all used nominal scales.

10.4 Results

To better understand the relationship between the different product development stages and sustainability and commitment to sustainability, we examined whether individuals associated with the four product development stages were different on the two sustainability measures. That is, commitment to sustainability and sustainable product development. In the following sections, we will report the results for each of the dimensions of sustainability.

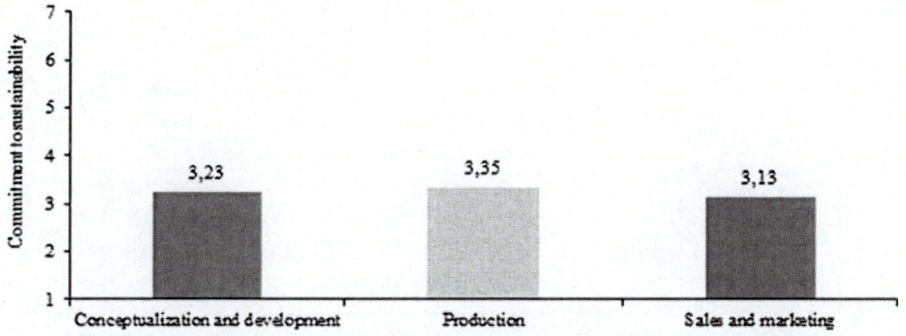

Fig. 10.2 Commitment to sustainability in the different stages of product development

10.4.1 Commitment to Sustainability

After reverse scaling the items for commitment to sustainability, we used an ANOVA to look for differences between the stages of new product development. The results of the ANOVA test did not show a significant effect of the product development stages on the participants' commitment to sustainability [$F\ (2, 49) = 1.00, p > 0.05$]. Thus, the null hypothesis of no differences between the means was not rejected. As can be seen in Fig. 10.2 the different stages were quite similar in mean levels of commitment with the mean of individuals working at the conceptualization and developmental level being 3,23, individuals from the production level being 3,35 and individuals from sales and marketing level at 3,13.

10.4.2 Sustainable Product Development

As the sustainable product development scale can be divided into three sub-scales, we start by looking at the scale as a whole before moving on to three dimensions (economic, environmental and social). All scales were reversed before running our analysis. Starting with the overall sustainable product development scale, we used an ANOVA test to look for differences between the stages of new product development. The results of the ANOVA test did not show a significant difference between the product development stages when examining their emphasis on overall sustainability ($F\ (2, 49) = 0.72, p > 0.05$). Thus, the null hypothesis of no differences between the means was not rejected. As can be seen in Fig. 10.3, the different stages were quite similar in mean levels of sustainability, demonstrating the mean of individuals working at the conceptualization and developmental level being 4,06, individuals from the production level being 3,87 and individuals from sales and marketing level at 3,98.

We next examined the sub-scales from the sustainable product development scale. For economic sustainability we found no difference between the three stages of

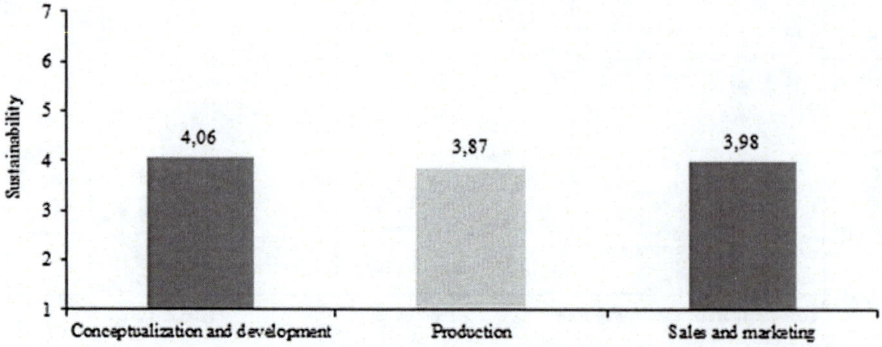

Fig. 10.3 Sustainability in the different stages of product development

product development (F (2, 49) = 1.44, $p > 0.05$). The same could be said for both environmental (F (2, 49) = 2.17, $p > 0.05$) and social sustainability (F (2, 49) = 0.748, $p > 0.05$). As can be seen in Fig. 10.4, the different stages had very similar sustainability levels for all three of the sub-scales of sustainability demonstrating the mean of commitment to the economic pillar of sustainability ranging from 3,53 from individuals within the conceptualization and development stage on the product development process, 308 form individuals within the production stage of product development and finally 3,42 from individuals within the sales and marketing stage. The mean of commitment to the environmental pillar of sustainability disseminates from a mean of 4,28 from individuals within the conceptualization and development stage, a mean of 4,30 from individuals within the production stage and a mean of 4,37 from individuals of the sales and marketing stage. Finally, the mean of commitment to the social pillar of sustainability ranged from the highest mean of 4,38 from individuals within the conceptualization and development stage, 4,23 from individuals

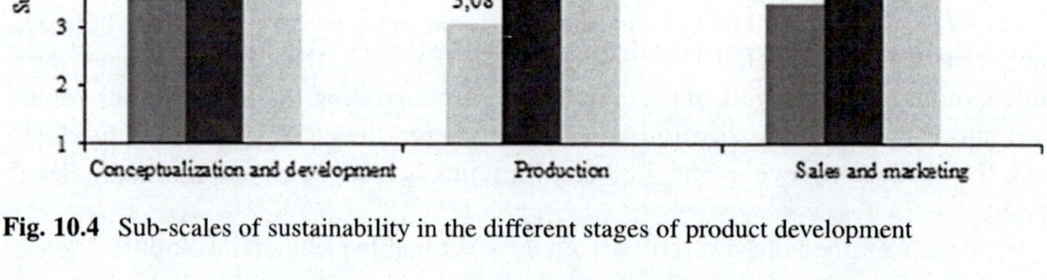

Fig. 10.4 Sub-scales of sustainability in the different stages of product development

within the production stage of product development and 4,17 from individuals within the sales and marketing stage.

As can be seen from these results we do not find that the stages of product development are associated with different levels of sustainability or commitment to sustainability. However, our data is limited to a small sample and should be considered exploratory.

10.5 Discussion and Conclusion

As consumers value towards products is shifting more and more towards sustainability and not financial value, organizations must commit their products to sustainability to meet their consumer's needs. This used the case study of the Icelandic marine sector to investigate the difference in individual commitment to sustainability between different phases of product development as well as contributing to literature about personal sustainability values. To our knowledge, this is the first study that investigates whether there exists a difference towards sustainability commitment in the different stages of product development within a particular economic sector.

As demonstrated in the findings chapter, the groups within product development behaved very similarly and had distinctive commitments towards sustainability. The statistical outcomes of this study indicate that professional sustainability commitments do not differ between stages of product development. Consequently, product developers should direct more of their attention to focusing on different sustainability attributes on different product developmental stages. One might wonder why individuals do not differ, if specialization between the stages is significant. It could also be that the complex and ever-changing environment in the Icelandic marine sector is a factor in the lack of differentiation between product development stages. In addition, the personal values of individuals towards sustainability seem to play an underlying role in their professional attributes to sustainability. Possibly individuals also find it difficult to relate to being in product development, even though their daily projects revealed otherwise.

Although future research could add to the sample size, since the result showed no significant difference between the two groups there is a chance that a new study could arrive at the same results. Namely, that the two groups have a small effective size and behave very similarly. The main disadvantage of using this specific sample is the generally small sample size within the sector.

Future research could also demonstrate a difference between the sustainability objectives of the organization as a whole and objectives of stages in product development. Qualitative research is also needed to demonstrate and give a deeper understanding of why the product development stages do not differ. Much remains to be understood concerning the sustainability of products and interactions with leaders and innovators in product development. It must not be forgotten that the leadership of an entrepreneur is likely to be fundamental to the level of sustainability of an innovation that is achieved. Given this, product developers and entrepreneurs may need to be educated to ensure a broader perspective than is common in many technical field todays, which suggests the need for more monitoring of their work situation to investigate the sustainability of new products arriving on the marketplace, and how this could be improved. More reflection needs to be given to describing what is a sustainable innovation and how it might be developed across different phases of product development. A sustainable solution must be long-lasting, environmentally responsible for the provider, i.e. the business, and society, including consumers.

10.6 Compliance with Ethical Standards

The authors declare that they complied with all guidelines given by the Science Ethics Committee at the University of Iceland. Informed consent was obtained from all participants and all data was anonymized. The research does not require ethics approval, as it mentioned in the waiver issued by the University of Iceland Research Ethics Committee.

Reference

1. Aguinis H (2019) Glavas A: on corporate social responsibility, sensemaking, and the search for meaningfulness through work. J Manag 45(3):1057–1086
2. Alcock I (2012) Measuring commitment to environmental sustainability: the development of a valid and reliable measure. Methodol Innov Online 7(2):13–26
3. Balderjahn I, Buerke A, Kirchgeorg M, Peyer M, Seegebarth B, Wiedmann K (2013) Consciousness for sustainable consumption: scale development and new insights in the economic dimension of consumers' sustainability. Acad Market Sci Rev 3:181–192
4. Bevilacqua M, Ciarapica FE, Giacchetta G (2007) Development of a sustainable product lifecycle in manufacturing firms: a case study. Int J Prod Res 45(18–19):4073–4098
5. Browning TR, Deyst JJ, Eppinger SD, Whitney DE (2002) Adding value in product development by creating information and reducing risk. IEEE Trans Eng Manag 49(4)
6. Chen AY, Lai S, Wen C, Journal S, Sep N, Chen Y (2012) The influence of green innovation performance on corporate advantage in Taiwan. J Bus 67(4):331–339
7. Cooper RG (2008) Perspective: the stage-gates idea-to-launch process—update, what's new, and nexgen Systems. J Prod Innov Manag 25:213–232
8. Cooper RG, Edgett SJ (2008) Maximizing productivity in product innovation: research technology. Manag 51(2):47–58
9. Dangelico RM, Pontrandolfo P, Pujari D (2013) Developing sustainable new products in the textile and upholstered furniture industries: role of external integrative capabilities. J Prod Innov Manag 30(4):642–658
10. Daily BF, Huang S (2001) Achieving sustainability through attention to human resource factors in environmental management. Int J Oper Prod Manag 21(12):1539–1552
11. Eizenberg E, Jabareen Y (2017) Social sustainability: a new conceptual framework. Sustain 9(68)
12. Florescu MS, Ceptureanu EG, Cruceru AF, Ceptureanu SI (2019) Sustainable supply chain management strategy influence on supply chain management functions in the oil and gas distribution industry. Energies 12(9):1632

13. González-Benito J, González-Benito O (2005) Environmental proactivity and business performance: an empirical analysis. Omega 33(1):1–15
14. Griffin A (1997) PDMA research on new product development practices: Updating trends and benchmarking best practices. J Prod Innov Manag 14(6):429–458
15. Hellström T (2007) Dimensions of environmentally sustainable innovation: the structure of eco-innovation concepts. Sustain Dev 15(3):148–159
16. Jabbour CJC, Jugend D, Jabbour ABLS, Gunasekaran A, Latan H (2015) Green product development and performance of Brazilian firms: measuring the role of human and technical aspects. J Clean Prod 87(15):442–451
17. Kara S, Ibbotson S, Kayis B (2014) Sustainable product development in practice: an international survey. J Manuf Technol Manag 25(6):848–872
18. Katsikeas CS, Leonidou CN, Zeriti A (2016) Eco-friendly product development strategy: antecedents, outcomes, and contingent effects. J Acad Market Sci 44:660–684
19. Krishnan V, Ulrich KT (2001) Product development decisions: a review of the literature. Manag Sci 47:1–21
20. Leonard-Barton D (1992) Core capabilities and core rigidities: a paradox in managing new product development. Strateg Manag J 13(S1):111–125
21. Luiz JVR, Jugend D, Jabbour CJC, Luiz OR, Souza FB (2016) Ecodesign field of research throughout the world: mapping the territory by using an evolutionary lens. Scientometr 109(1):241–259
22. Luque-Vilchez M, Mesa-Perez E, Husillos J, Larrinaga C (2019) The influence of pro-environmental managers' personal values on environmental disclosure: the mediating role of the environmental organizational structure. Sustain Account Manag Policy J 10(1):41–61
23. Olsen EM, Walker OC, Ruekert RW (1995) Organizing for effective new product development: the moderating role of product innovativeness. J Market 59(1):48–62
24. Pinheiro MAP, Seles BMRP, Fiorini PC, Jugend D, Jabbour ABLS, Silva HMR, Latan H (2019) The role of new product development in underpinning the circular economy: a systematic review and integrative framework. Manag Decis 57(4):840–862
25. Pujari D (2006) Eco-innovation and new product development: understanding the influences on market performance. Technovation 26(1):76–85
26. Sanyé-Mengual E, Pérez-López P, González-García S, Lozano RG, Feijoo G, Moreira MT (2014) Eco-designing the use phase of products in sustainable manufacturing. J Ind Ecol 18(4):545–557
27. Shao J, Ünal E (2019) What do consumers value more in green purchasing? Assessing the sustainability practices from demand side of business. J Clean Prod 209:1473–1483
28. Sun M, Rydh CJ, Kaebernick H (2003) Material grouping for simplified product life cycle assessment. J Sustain Prod Des 3:45–48
29. Trimi S, Berbegal-Mirabent J (2012) Business model innovation in entrepreneurship. Intern Entrep Manag J 8:449–465
30. Veryzer RW (1998) Discontinuous innovation and the new product development process. J Prod Innov Manag 15(4):304–321
31. Vinodh S, Rathod G (2010) Integration of ECQFD and LCA for sustainable product design. J Clean Prod 18(8):833–842
32. Vinodh S, Rathod G (2011) Application of ECQFD for enabling environmentally conscious design and sustainable development in an electric vehicle. Clean Technol Environ Policy 13(2)
33. Wenzel H, Hauschild M, Alting L (2001) Environmental assessment of products. Chapman and Hall, London
34. World Commission on Environment and Development (1987) Our Common Future (The Brundtland Commission). Oxford University Press, Oxford

Chapter 11
Inter-Organizational Co-Creation: An Approach to Support Energy Transition Projects

Afshin Jalali Sohi, Maryam R. Nezami, Hans Bakker, and Marcel Hertogh

Abstract Societal challenges such as climate change and inefficiency of energy systems more and more crave for a sustainable environment. Research proved that restructuring energy systems into more sustainable forms, called "Energy Transition", has faced challenges. How to deal with these challenges requires the co-creation between various actors with multiple disciplines, expertise, and perspectives from different organizations. The research question to be answered here is whether co-creation helps the interaction between different actors in an inter-organizational project for the sake of better project results. By doing case study research in the Netherlands, an example of co-creation project in its front-end phase was investigated regarding the interaction among different actors involved in the project. The research revealed that in the case of a co-creation project the multiple actors collaborate across organizational boundaries in order to unite. This leads to a better solution-finding approach. Openness, trust and respect are valued more in co-creation. Moreover, the project team is better integrated to work towards a shared interest which are social benefits. The co-creation facilitated the data-sharing among the key actors in the project which traditionally is influenced highly by the mother-organizations' culture. Further research will investigate the transition in organizations to support the co-creation approach.

Keywords Energy transition · Co-creation · Inter-organizational projects · Collaboration

A. Jalali Sohi (✉) · M. R. Nezami · H. Bakker · M. Hertogh
Delft University of Technology, Stevinweg 1, Delft 2628 CN, The Netherlands
e-mail: afshin.jalalisohi@gmail.com

M. R. Nezami
e-mail: M.Rikhtegarnezami@tudelft.nl

H. Bakker
e-mail: H.L.M.Bakker@tudelft.nl

M. Hertogh
e-mail: M.J.C.M.Hertogh@tudelft.nl

R. Cuevas et al. (eds.), *Research on Project, Programme and Portfolio Management*, Lecture Notes in Management and Industrial Engineering,
https://doi.org/10.1007/978-3-030-60139-3_11

11.1 Introduction

The Dutch ministry of Economic Affairs and Climate Policy in 2016 had the ambition of transitioning towards a low-carbon energy economy that is safe, reliable and affordable. The strategy for the energy transition includes focusing on reduction of CO_2 emissions, supporting innovation and seizing economic opportunities, integrating the energy transition into spatial planning, introduction of energy functionalities, and last but not least to further develop the strategy through "energy dialogue with the public, businesses, NGOs and government authorities". The coalition agreement of the Dutch cabinet in 2017 set the national strategy on 49% CO_2 emission reduction by 2020 while the international strategy aims at 55% CO_2 emission reduction [42]. This difference between the objectives of the Netherlands compared to Europe ignites the necessity of speeding up the energy transition in the Netherlands to reach the European Union goals.

The Dutch National Research Agenda as a source of inspiration for scientific developments for the direction of societal challenges for society in future offers 140 overarching scientific questions in 16 divers directions [11]. In line with the move towards energy transition one of those 16 directions is dedicated to energy transition. The purpose of the research is to build a sustainable and secure energy supply and a strong, green knowledge-based economy. It is believed for such an energy transition an integrated approach is required which addresses technical, social, economic, legal and spatial challenges that allow excellent building blocks to be implemented quickly and on a large scale [11].

The energy transition may have effects on many businesses and innovation strategies both locally and globally. It certainly questions the survival of companies like Shell which works on extraction and an executive refinement of fossil fuel if the world transitions to lower carbon energy production and consumption. Chad Hollidays from Shell claims that Shell has a flexible strategy to keep in step with the changes in the energy system [37]. It can be argued that all the businesses which rely on fossil fuel feel the societal urgency for sustainability purposes.

Although energy transition is a goal, the performance of those projects are not promising based on research [22, 33, 45]. It was understood that energy transition is a global goal. However, it has faced challenges in the countries who start moving to reach those goals. For example, German renewable energy cooperatives struggled as market collapsed as cited by Buchsbaum [8]. He claimed that the industry is paralyzed by the downwards pressure on onshore wind energy expansion. Another example is in The Netherlands. As reported by Netherlands Energy Research Alliance (NERA) energy transition has faced a number of challenges such as clean and flexible industry, intelligent energy systems, the Dutch transition in a global context and dealing with variation [30]. NERA Stated that, in order to support the energy transition, the cooperation in the market is required [2]. This cooperation can be realized between the policymakers, infrastructure asset owners and the cooperating countries in the energy transition agreement. However, the facts and figures reported by the European Commission show that the international transition is lagging behind of what has been targeted as interconnection capacity [13]. Therefore the cooperation between the actors in energy transition plays an important role in goal achievement of the energy transition.

By studying the challenges of energy transition in Germany, Pegels et al. [33] stated that energy transition needs a formation of transformative alliances to be successful. In their viewpoint alliances will need to go beyond conventional boundaries. They believe that success of a transformation alliance is dependent on a platform of employment, competitiveness and innovation.

Insufficient supporting infrastructures is one of the main challenges energy transition has faced [10]. When it comes to realization of infrastructure, there is no single organization or authority responsible but numerous actors are involved. This becomes even more important when it comes to international energy transition. In a study about electricity production (high voltage direct current) in energy transition Pierri et al. [34] cited three categories of challenges: technical, economical and social, and environmental challenges. They propose that a solution to solve the challenges and achieve an agreement is the strict collaboration between existing associations and involved actors like energy companies.

Blanchet [5] studied the governance of energy transition projects in Berlin. He mentioned that energy transition will need transition management in regard to socio-technical transitions. Based on the research he claimed that the inclusion of local stakeholders, creating common interest through framing, familiarizing the public with the issue through forums and advertising are the strategies in energy initiatives. Heiskanen et al. [19] studied the emergence of sustainable energy transition from the perspective of two main energy companies in Finland. They found that new forms of collaboration are emerging within the coalition for energy transition. This new form of collaboration requires the combination of actors working together for a joint purpose.

It became clear that energy transition is "a must" for societies for sustainability purposes. It is also evident that the road to achieve energy transition is not as smooth as it was hoped for and there are challenges to achieve the set objectives. One main challenge is the collaboration of different actors in energy transition projects. Talking about "different actors" the focus is on inter-organizational collaboration where different organizations come together for a shared interest, in this case energy transition. Collaboration has different forms and co-creation is one of them. In this research a closer look has been taken to an example of a co-creation energy transition project. The objective of the research is to study how different organizations in an inter-organizational setting co-create to achieve a set goal.

In the paper Sect. 11.2 elaborates on the literature review regarding the concept of collaboration. Section 11.3 elaborates on the research methodology. The research results are presented in Sect. 11.4. Sections 11.5 and 11.6, respectively, cover the discussion and conclusion of the research.

11.2 Literature Review

In Sect. 11.1 it was argued that energy transition is a necessity for the modern society to overcome some societal challenges regarding sustainability. It was also mentioned that energy transition is not possible without a good collaboration between different actors such as client organizations, energy companies and NGOs among others. This section aims at providing a background study on collaboration. Further, this section explains what it is meant by inter-organizational projects.

Collaboration is an universal activity in modern societies and is recognized as a promising approach to address organizational and societal problems [17]. The word of collaboration is originally derived from the Latin word *collaborare* that means "work with" [18]. Different scholars provide different definitions of collaboration. Collaboration has been defined as the interdependent work of people together to achieve a greater interest and goal than they can attain individually [25]. According to Lu et al. [25], collaboration means: "*any effort to collaborate to exchange information, ideas or useful resources necessary to create a shared understanding for a common and creative purpose*". Based on the definition of Bryson et al. [7], collaboration is "*the process intended to foster sharing that is necessary among involved or affected groups or organisations in order to achieve the collective gains or minimise the losses*". Gray [16] proposed the definition of collaboration in her book as a process through which "*parties who see different aspects of a problem can constructively explore their differences and search for solutions that go beyond their own limited vision of what is possible*".

Various individuals are involved in collaboration and joint their efforts in order to obtain mutually desirable goals [6]. Collaboration is considered in the literature as an umbrella term for alliancing, networking, joint ventures and partnering [20].

There are other terms such as cooperation and coordination that sometimes interchangeably have been used by practitioners or in literature. McNamara [27] identified the differences between collaboration and other forms of working together—such as cooperation and coordination. According to him, while cooperation put the emphasis on individuals, coordination acknowledges the relationships in team and collaboration is centred by the relationships.

Collaboration creates long-term relationships through the involvement of participants which lead to greater innovation to achieve common goals [23]. In addition, close collaboration results in decreasing lead-time and improve quality in design and development [28]. Azari and Kim [1] recognized the parameters that have an effect on the quality of collaboration in the construction projects as accountability, commitment, communication, compatibility, timely involvement, joint operations, mutual respect and trust. The council of administrators of special education make a list of recommendation for collaboration [23]: listen to each, fully understand alternative opinions and perspectives, find and work from common ground, try to elucidate concerns and expectations, respect other people's differences, make use of each other's strengths and expertise, discover alternative ideas or perspectives and work together to attain the best possible solutions.

The term collaboration is mainly used when different organizations/parties work together to achieve an objective together which is not achievable without collaboration. Based on who those parties are in any form of collaboration, different researchers found different factors of collaboration.

Previous studies on this subject have mainly concentrated on collaboration and the relationship between client and contractors [39], between the owner, design teams and contractors [38] and collaboration in project-based supply chains [24, 26]. However, inter-organizational collaboration in multi-client projects is less addressed in the literature. Hence, the focus of this research is on factors of collaboration in the form of co-creation in multi-client inter-organizational projects.

According to Jones et al. [21] an inter-organizational project: *"involves two or more organisational actors from distinct organisations working jointly to create a tangible product/service in a limited period of time"*. The main difference between an inter-organizational project and other forms of collaboration is the fact that projects are temporary [21] but not any form of collaboration is temporary. In case of inter-organizational projects, the project organization dissolves after achieving common goals. A wide range of industries such as advertising [15], construction [12, 46], biotechnology [36], computer [14], financial services [35] have applied this kind of joint working of various organizations. However, there are few frameworks of understanding the different kinds of inter-organizational projects [21].

By reviewing the literature it was concluded that collaboration can be generally defined as a working condition in which different actors come together for achieving a shared interest/goal while the achievement of the goal is impossible individually. It was discussed that collaboration is different than cooperation and coordination. Collaboration can be applied in different forms, however, at the end the purpose of collaboration is achievement of shared goals. Collaboration can be among any two or more parties such as client-contractor, contractor-subcontractor, contractor-supplier and many others. In this research the focus is on collaboration between different client organizations for energy transition purposes.

11.3 Research Methodology

The objective of the research is to study how different client organizations in an inter-organizational setting collaborate in the form of co-creation to achieve a set goal. The research question to be answered here is whether co-creation, as a form of collaboration, helps the interaction between different actors in an inter-organizational project for better project results. In order to answer the research question, a case study research [47] is performed. To gather information about the project two approaches were taken: document study and semi-structured interviews with the key roles in the project and parent organizations.

The interviews intended to get an insight about the soft factors of collaboration which were not possible to get from the document study. For sensitivity reasons the name of the project and interviewees were kept anonymous.

The project is about infrastructural objects (such as locks and sluices) in the Netherlands that are reaching end of their life cycle. The first possible solution is to replace these objects by new ones which will be designed again for another 100 years. However, in 100 years' time the same issue will raise that is faced now: replacement of objects. Therefore, there should be a more innovative solution which answers nowadays societal challenges. All these objects are owned, constructed, operated and maintained by different organizations. However, there are always one or two organization which is/are known as the asset owner. Most often these infrastructures are interconnected, meaning that replacement/renovation of one object has consequences for other objects as well. To come up with possible scenarios for the replacement of those objects the asset owners need to come together and collaborate to reach a better result.

Although the replacement of those objects is the current issue, the obligation for moving towards sustainable energy and CO_2 emission reduction asks for a more integrated approach towards energy transition. Hence, one of the possible scenarios would be the integration of replacement of those objects with energy transition projects.

The project was initiated with an open question to the public. So any interested party could join and brainstorm in finding a solution for this project. The fact that it was an open question to the public resulted in an integration of the project with energy transition goals. The initial idea was finding a solution for the replacement of those infrastructural objects. However, energy companies showed interest in the project which lead to integration of the project with energy transition purposes. Since the number of those infrastructural objects is big, it was decided to start with a pilot location.

The project is at its front-end phase. The front-end phase itself can be divided into sub-phases. In this case, the front-end was split into six sub-phases. The decision of splitting the phase into six sub-phases was made to extend the process with more clear stage gates. The first phase of the project was mainly brainstorming for ideas to solve the current issue (replacement of infrastructural objects which are at the end of their lifecycle). This phase was the opportunity framing phase where different scenarios were developed as possible solutions. The integration of energy transition into the basic project purpose was initiated in this phase. In phase two the project team focused on less scenarios (funnelling the ideas) from the perspective of the asset owner and the energy company. The output of phase two was the business case document and technical design at a high level. Therefore phase two focuses on answering the question on generic and local level (technical, legislation, rules, etc.) in a form of a business case. Phase three is about making a plan for specific location (pilot location). Phase four is about preparing for the project with detail design. Phase five is then executing it and phase 6 is evaluating and turning it into a form of standard which can be applied and repeated in other locations.

This project fits well with the definition of an inter-organizational project since there is more than one organization involved (the asset owner and the energy company). Moreover, there was a common goal for the involved parties in the project.

Table 11.1 Interviewees functions

	Function	Project/parent organization
Interviewee 1	Investment manager	Project
Interviewee 2	Project leader	Project
Interviewee 3	Process manager (secratory)	Project
Interviewee 4	Senior adviser	Project
Interviewee 5	Adviser	Parent organization
Interviewee 6	Project manager innovation (energy transition)	Parent organization
Interviewee 7	Strategy director	Parent organization

The common goal here was the social benefit. The emphasis should be put on the total social benefit while not threatening each other's interests.

In total seven interviews were performed: 4 at project level and 3 at the parent organizations. Table 11.1 presents the interviewees' profiles. All the interviews were transcribed. Then by coding the relevant key words (collaboration, cooperation, co-creation, team integration, project team) the relevant parts were extracted. By qualitatively analysing the data, the perception of collaboration and the factors of collaboration were extracted. By doing a desk research on project documents also the factors of collaboration (in this case co-creation) were extracted. After compiling the results of interviews and desk research the final list of collaboration factors was concluded. Section 11.4 elaborates on the research results.

11.4 Results

At first instance, it is important to explore what the project team perceives as collaboration. Interviewees were asked to provide their definition of collaboration (Table 11.2).

Some general dimensions of collaboration mentioned by the interviewees are (1) setting a common goal (2) working together (3) shared interest over individual interests (4) horizontal network structure and (5) openness in communication. While practitioners perceived collaboration as working together towards a common goal, they perceived co-creation as creating a new thing by working together. Thus the difference between collaboration and co-creation is mainly about creativity and innovativeness in creating something new.

Although most of the interviewees interpret co-creation as a form of collaboration, still some believe that co-creation goes beyond the collaboration. Interviewee 3 perceives collaboration different than co-creation by pinpointing it to the contract: binding the parties, hierarchy and expectations from each other. She believes that the emphasis in co-creation is on creating something new which makes it different than collaboration, although "working together" is a pre-condition for both collaboration and co-creation. Innovativeness and creativity is mentioned often as the keys

Table 11.2 Definition of collaboration by interviewees

Interviewee	Definition of collaboration/co-creation
Interviewee 1	Collaboration in this field is co-creation. Collaboration is having discussions about the way to co-create with each other. The output of discussion will be published
Interviewee 2	Collaboration is working together, not putting anybody's interest first, but jointly looking for shared interest
Interviewee 3	Co-creation is to create something new in a group working together with no boss (hierarchy), expectations and contract binding
Interviewee 4	Collaboration is about being open in communication
Interviewee 5	Collaboration is having a mutual goal and try to achieve it together
Interviewee 6	Collaboration is being able to take and to give in the search of reaching the common goal
Interviewee 7	Collaboration is about "why" and not "how". Recognition of public values and being clear how to achieve it

for co-creation. The project leader from the energy company believed that the word co-creation is the same as collaboration and people use it if they want to give collaboration more flavour. However, after working in a co-creation team, he mentioned that co-creation might have a particular meaning compared to collaboration although to him both concepts are the same. Interview 2 mentioned that in collaboration you need to form a team which it has a kind of identity. That identity keeps the team together towards a goal rather than a group of people from different organizations sitting at the opposite side of the table.

In the co-creation of an inter-organizational project the project team set some basic principles of co-creation. The purpose of doing so was to make it clear for the participants in the project what to expect from co-creation. The principles helped the team to have a clear picture of the process. Table 11.3 presents those basic principles.

Table 11.3 The basic principles of co-creation in the case study project

Basic principles for co-creation in general
Full Transparency
Joint search for joint steps
Respecting each other
Defining the playing field
Thinking of possibilities
Setting clear goals
Ensure the direction; control of the process
Providing room/space for everyone
Take care of each other
Knowing and respecting each other's interests
Apply network organization rules

Table 11.4 Positive and negative aspects of co-creation recognized by the project team

Positive aspects	Negative aspects
Setting multiple challenging goals	Assuming that the outcome for the asset owner is fixed
The right of the initiators	Presentation does not reflect the reality
Sharing the knowledge and skills that are needed, in a timely manner in the process	Keeping only the related content for the asset owner (exclusion)
Dedicating time and space	Having a tunnel vision in the process
Give each other incentives	Lack of space
Make the interests clear	Thinking in problems
Helping each other in the thinking process	Hidden agendas
Being able to make personal contribution	No equivalence of values
Achieving success, progress and energy (in process)	Abuse
Working towards social values	Continuous process discussions
People, Planet and profit	No added value
Motivation, intrinsic values	

As can be seen in the table, the basic principles are meant for better alignment of the team and understanding of co-creation. The emphasis was put on equality in the team by, for example, "knowing and respecting each other", "network organization" and "goal setting".

Those basic principles of co-creation could lead to some specific positive or negative aspects as recognized by the project team at the early stage of the project (phase 1). Table 11.4 elaborates on the identified positive and negative aspects of the co-creation.

By comparing the identified positive and negative aspects of co-creation it can be argued that the positive aspects are mainly about the integrity of the team as a single unit focusing on a shared interest. However, the negative aspects stress the tunnel vision and the situation where there is no essence of collaboration and innovativeness because of being restricted to presumptions. Those presumptions hinder the collaboration. For example, "hidden agendas" is a threat to openness and trust. "Exclusion" of any parts in the project without being open about it also threatens the trust and consequently the collaboration. Thus, the negative aspects should be avoided and recognized in the early stages of the project in order to achieve a successful collaboration.

Apart from the basic principles of co-creation, the interviews resulted in gathering a list of factors which influence the collaboration. Table 11.5 presents the factors of collaboration.

By analysing the factors mentioned by the interviewees for collaboration it can be said that most of the factors contribute to soft aspects within project team such as trust, sympathy, respect, team spirit, top management support, people over instructions, commitment and equality among the actors in an inter-organizational project. The second category of factors is related to project management like setting a clear goal,

Table 11.5 Factors of collaboration in inter-organization project identified by interviewees

	Factors of collaboration	Frequency of being mentioned by interviewees
1	Interested and enthusiastic people	2
2	Trust	2
3	Curiosity	2
4	Common goal	2
5	Working together	2
6	Common ground	2
7	Shared understanding from each other (get to know each other)	2
8	Support from the top management	2
9	High team spirits (having fun)	2
10	Make an integral business case	1
11	Sympathy	1
12	Respect	1
13	People over instructions	1
14	No monetary influence	1
15	Joint agenda	1
16	Commitment	1
17	Stable team	1
18	No restrict division of functions	1
19	Open knowledge sharing	1
20	Horizontal organization (no hierarchy)	1
21	Clarity of the goal	1
22	Clear communication	1
23	Early involvement of stakeholders (depending on every stage of the project)	1
24	Contractual agreement	1
25	Loose of ownership	1
26	Iterative progress	1
27	Equality (everybody has a voice and should be heard)	1

joint agenda's, integral business case and contractual agreements. The third group of factors is those which are related to processes such as clear communication, open knowledge sharing, and early involvement of stakeholders. The last identified group of factors are those which contribute to team structure such as stable team and horizontal organization.

One of the interviewees mentioned that in co-creation we are not anymore parties but people who sit together around the table as a team. This observation reflects on the integration of team rather than separated organizations around the table. Another interviewee supports the idea of integrated team by mentioning that in a co-creation project environment everybody is the same in the team. This means that there is no priority/importance given to any sort of hierarchy in the project organization. That is why the horizontal project organization is mentioned as a factor of inter-organizational collaboration.

One of the factors of inter-organizational collaboration mentioned by interviewees was "getting to know each other". One of the team members in reflecting about the co-creation process stated that: "I am positive about the fact that the project team made so much progress although it was made of people who didn't know each other before. The vision of others helped shaping our own ideas". This means that even though the project team didn't know each other before, the co-creation process enhanced the collaboration among them which resulted in good project progress. Interviewee 2 mentioned: "you need to get to know each other in order to see what kind of plans fit both interests". He believed a common knowledge about the project is required for better understating of each other.

The interaction among team members is reflected in project documents. There was an observation by a team member that as learning point we should listen better to each other. Another team member supports this idea by mentioning that there should be a balance between listening and speaking.

"Transparency" was mentioned as a factor of inter-organization collaboration. Interviewee 3 supports this factor by saying that "in co-creation everybody sits in the team for the public good. In this case nobody was allowed to keep information behind".

Apart from the positive side of co-creation there were also negative aspects observed by the team. A team member said that there was a "structured chaos" way of doing things in some meetings. When different people with different backgrounds come together without any structure, only the last five minutes would be spend on the core subject. This observation by the team member is interpreted by the researchers as inefficiency of meetings in co-creation. Regarding the chaos another team member said: "as long as chaos is intrinsically motivated, it will lead to success". Therefore, it can be argued that there is some structure needed in a co-creation process to increase the efficiency. Another negative aspect observed by an interviewee was the existence of "group thinking". Interviewee 2 mentioned that "group thinking" is a drawback of group meetings until somebody starts criticizing the idea. Then everybody starts questioning the progress.

Regarding the importance of having a stable team for a co-creation project one of the interviewees mentioned: "the team should be stable. You can't have people running in and out all the time because you have to build knowledge". The importance of knowledge remaining in the team by establishing a stable team is also highlighted in literature about Agile project management [3].

11.5 Discussion

The first topic which was explored in this research was the perception of collaboration. It was revealed that almost all interviewees perceived collaboration as working together to achieve a common goal. Interviewee 2 mentioned that to achieve the social good (common goal) the parties should always comprise their benefits. This highlights the importance of interaction among the involved actors in the inter-organizational collaboration project.

Explaining the perceptions, interviewees mentioned "contract" as a factor which can affect collaboration. Collaboration can be realized in form of co-creation. The first observation made was that although all interviewees have a clear picture of co-creation, not all of them see the collaboration and co-creation as the same concept. For example, interviewee 3 believed that in collaboration there is a contract which plays an important role while in co-creation there is no contractual agreement. Literature also supports the fact that collaboration is affected by contractual agreements [40]. Co-creation in literature is when the customer also takes a role in creating the value [32]. According to research [32] the development of relationships between two actors is based on interaction and dialogue. Mele [29] states that value should be co-created rather than created by the service provider for the customer. By a literature scan it was concluded that most of the literature on co-creation has looked into the relationship between customer (client) and providers (suppliers) [29, 32, 44]. This research concludes that co-creation hasn't been studied in the context of multi-client inter-organizational projects.

Collaboration has been studied by different researchers [4, 9, 31, 39, 41, 43, 48]. However, collaboration in multi-client inter-organization projects has not received much attention among scholars. Hence, in this research the secondary objective was to explore the factors of collaboration which are recognized by practitioners in case of an inter-organizational project. in total 27 factors of collaboration were identified. Most of the factors contribute to soft factors of collaboration. This can strengthen the idea that the motives of collaboration are soft aspects of project management and not hard aspects such as contracting. Also it was evident that team related factors such as "transparency", "respect", "trust", "enthusiast people", "commitment" and "high team spirit" get the highest attention among the practitioners.

Comparing the identified factors in inter-organizational collaboration in this research with the factors of other forms of collaboration in literature [39] it can be concluded that factors of collaboration are mostly the same, no matter what form of collaboration it is and between which two actors in the project. Suprapto et al. [39] identified different categories of collaboration factors between client and contractor such as relational attitudes, collaborative practices, joint capabilities and teamwork quality. Relational attitudes include factors such as management support and commitment, valuing each other's interests. Collaborative practices include items such as formal integrated project team, goal setting, joint decision-making among others. Joint capabilities are about competences and experiences of actors in collaborative relationship. Teamwork quality is about factors such as communication, coordination and cohesion, affective trust, aligned effort and balanced contribution. By comparing these collaboration factors and the identified factors in this research it can be concluded that the joint capability items were not recognized by practitioners regarding the collaboration in multi-client inter-organizational projects. The rest of factors show a huge overlap.

By studying the collaboration in a multi-client inter-organizational project a few managerial implications can be drawn from this research. First of all, collaboration is highly influenced by the shared interest of actors rather than contract. This helps practitioners by putting more emphasis on the people side of the project rather than contracting. Secondly, the recognition of collaboration factors in the context of multi-client inter-organizational projects helps practitioners empower their collaborative relationships to reach mutual goals for the society.

The scientific contribution of this research is bridging the gap in literature regarding the perception and the factors of collaboration in a multi-client inter-organizational context. For the successful delivery of inter-organizational projects the collaboration between different actors is required. However, the collaboration between multiple client organizations has not been studied thoroughly before. This collaboration is of a bigger importance in case of energy transition projects where the achievement of project goals goes beyond the contribution of a single client organization.

11.6 Conclusion

Literature suggests that new forms of collaboration are required for energy transition projects [19]. In most cases the energy transition projects are not possible to be executed without inter-organizational collaboration. It is inherently inter-organizational since the scale of complexity of energy transition projects is on one hand beyond the capabilities of single organizations and on the other side the energy transition projects are most often interconnected infrastructure projects where different organizations are involved. Therefore one of the characteristics of these types of innovative projects is being inter-organizational. It was also evident that energy transition projects face challenges. Overcoming the managerial and social challenges is impossible without a good collaboration between the actors of energy transition projects. In this research a closer look was taken to a case study where energy transition goals were integrated with the initial goals of project (replacement of old infrastructure projects). In order to realize a good collaboration, this project was run as a co-creation project which is recognized as a form of collaboration. The research revealed that co-creation is a form of collaboration where the focus is more on innovativeness and creativity for finding a solution for a societal challenge. Therefore the factors of collaboration apply to co-creation context such as “shared interest”, “common goal”, “transparency”, “openness”, “respect” and “trust” among others. It was evident that soft factors of collaboration play important roles in the success of collaboration as they were more recognized by the practitioners than hard factors such as contractual agreements. The integration of project the team is identified as one of the main characteristics of co-creation. The existence of the identified collaboration factors does not guarantee the existence of collaboration. However, the nonexistence of those factors hinders the existence of collaboration in inter-organizational projects. This leads to the first managerial implication of this research which is recognition and implementation of collaboration factors in the context of inter-organizational energy transition projects. The scientific contribution of this research is bridging the gap in literature regarding the concept of multi-client inter-organizational collaboration. Further research can focus on the effect of identified inter-organizational collaboration factors on project success. Also this research can be the starting point for further research investigation in transitions in organizations to support the co-creation approach.

Acknowledgements This research is funded by Netherlands Organization for Scientific Research (NWO).

Compliance with Ethical Standards

This research has been performed under the permission of Delft University of Technology (reference number 439.16.804) and it is funded by Netherlands Organization for Scientific research (NWO). The research has been done with respect to Delft University of Technology regulations on research Integrity code of conduct. The ethical approval of the research has been issued and signed by Dr. Ir. Udo Pesch as the chair of Human Research Ethical Committee at the Delft University of Technology.

All the interviewees participated in this research were informed about the purpose of the research and publication. They all agreed on participation. All the data such as the names of interviewees, projects and involved organizations for this research are anonymized. Moreover, the interviewees were informed about the anonymization of the data prior to their participation in the research.

References

1. Azari R, Kim Y-W (2015) Integration evaluation framework for integrated design teams of green buildings: development and validation. J Manag Eng 32(3):04015053
2. Baron R (2016) Energy transition after the paris agreement: policy and corporate challenges WindEurope Summit: Hamburg
3. Beck K, Beedle M, Bennekum AV, Cockburn A, Cunningham W, Fowler M, Grenning J, Highsmith J, Hunt A, Jeffries R, Kern J, Marick B, Martin RC, Mellor S, Schwaber K, Sutherland J, Thomas D (2001) Manifesto for Agile Software Development. Available from: https://agilemanifesto.org/, 01-03-2014
4. Binder J (2016) Global project management: communication, collaboration and management across borders. Routledge
5. Blanchet T (2015) Struggle over energy transition in Berlin: How do grassroots initiatives affect local energy policy-making? Energy Policy 78:246–254
6. Briggs R, Kolfschoten G, de Vreede G-J, Dean D (2006) Defining key concepts for collaboration engineering. AMCIS 2006 Proc, 17
7. Bryson J, Eden C (1995) Addressing public problems through collaboration: the role of 'not-goals' and the problem of assessing accountability for their achievement'. In: 2nd International Workshop on Multi-organisational Partnerships: working together across organisational boundaries, Glasgow

8. Buchsbaum LM (2019) German renewable energy cooperatives struggle as markets collapse. Energy Transition: the global energiewende; Available from: https://energytransition.org/2019/06/german-renewable-energy-cooperatives-struggle-as-markets-collapse/
9. Caniëls MCJ, Chiocchio F, van Loon NPAA (2019) Collaboration in project teams: The role of mastery and performance climates. Int J Project Manag 37(1)::1–13
10. Dorian JP, Franssen HT, Simbeck DR (2006) Global challenges in energy. Energy Policy 34(15):1984–1991
11. Dutch-National-Research-Agenda, Portfolio for Research and Innovation (2018)
12. Eccles RG (1982) The quasifirm in the construction industry. J Econ Behav Organ 2(4):335–357
13. European-Commission, A framework strategy for a resilient energy union with a forward-looking climate change policy. Communication from the Commission to the European Parliament, the Council, the European Economic and Social Committee, the Committee of the Regions and the European Investment Bank, COM. 80, (2015)
14. Gomes-Casseres B (1994) Group versus group: how alliance networks compete. Harvard Busin Rev 72(4):62–66
15. Grabher G (2002) The project ecology of advertising: tasks, talents and teams. Regional Stud 36(3):245–262
16. Gray B (1989) Collaborating: Finding common ground for multiparty problems
17. Gray B, Wood DJ (1991) Collaborative alliances: moving from practice to theory. J Appl Behav Sci 27(1):3–22
18. Harper D (2001) Online etymology dictionary
19. Heiskanen E, Apajalahti E-L, Matschoss K, Lovio R (2018) Incumbent energy companies navigating energy transitions: strategic action or bricolage? Environ Innov Soc Trans 28:57–69
20. Hughes D, Williams T, Ren Z (2012) Differing perspectives on collaboration in construction. Constr Innov 12(3):355–368
21. Jones C, Lichtenstein BB (2008) Temporary inter-organizational projects, in The Oxford handbook of inter-organizational relations
22. Kaldellis JK, Kapsali M (2013) Shifting towards offshore wind energy—recent activity and future development. Energy Policy 53:136–148
23. Kinsella-Meier MA, Gala NM (2016) Collaboration: definitions and explorations of an essential partnership. Odyssey: New Directions in Deaf Education 17: p 4–9 (2016)
24. Koolwijk JSJ, van Oel CJ, Wamelink JWF, Vrijhoef R (2018) Collaboration and integration in project-based supply chains in the construction industry. J Manag Eng 34(3):04018001
25. Lu SCY, Elmaraghy W, Schuh G, Wilhelm R (2007) A scientific foundation of collaborative engineering. CIRP Ann Manuf Technol 56(2):605–634
26. McIvor R, Humphreys P, Cadden T (2006) Supplier involvement in product development in the electronics industry: a case study. J Eng Tech Manage 23(4):374–397
27. McNamara M (2012) Starting to untangle the web of cooperation, coordination, and collaboration: a framework for public managers. Int J Public Administr 35(6):389–401
28. Meijer BR, Voûte JH, Tomiyama T (2004) Communicating context and strategy for collaborative design in networks and corporations. In: Methods and Tools for Co-operative and Integrated Design. Springer, pp 363–374
29. Mele C (2011) Conflicts and value co-creation in project networks. Ind Mark Manage 40(8):1377–1385
30. NERA. Netherlands Energy Reaserch Agenda, Ten challenges for the energy transition. 2019; Available from: https://www.nera.nl/ten-challenges-for-the-energy-transition/
31. Oraee M, Hosseini MR, Papadonikolaki E, Palliyaguru R, Arashpour M (2017) Collaboration in BIM-based construction networks: a bibliometric-qualitative literature review. Int J Project Manage 35(7):1288–1301
32. Payne AF, Storbacka K, Frow P (2008) Managing the co-creation of value. J Acad Mark Sci 36(1):83–96
33. Pegels A, Lütkenhorst W (2014) Is Germany's energy transition a case of successful green industrial policy? Contrasting wind and solar PV. Energy Policy 74:522–534

34. Pierri E, Binder O, Hemdan NGA, Kurrat M (2017) Challenges and opportunities for a European HVDC grid. Renew Sustain Energy Rev 70:427–456
35. Podolny JM (1993) A status-based model of market competition. Am J Sociol 98(4):829–872
36. Powell WW, Koput KW, Smith-Doerr L (1996) Interorganizational collaboration and the locus of innovation: Networks of learning in biotechnology. Administr Sci Quart, 116–145
37. Shell, Shell Energy Transition Report (2019)
38. Smyth, Pryke S (2009) Collaborative relationships in construction: developing frameworks and networks. Wiley
39. Suprapto M, Bakker HLM, Mooi HG (2015) Relational factors in owner–contractor collaboration: the mediating role of teamworking. Int J Project Manage 33(6):1347–1363
40. Suprapto M, Bakker HLM, Mooi HG, Hertogh MJCM (2016) How do contract types and incentives matter to project performance? Int J Project Manage 34(6):1071–1087
41. Um K-H, Kim S-M (2018) Collaboration and opportunism as mediators of the relationship between NPD project uncertainty and NPD project performance. Int J Project Manage 36(4):659–672
42. Van 't Hof W (2018) Energy transition in the Ntherlands—phasing out of gas. Ministry of Ecoonomic Affairs and Climate Policy The Hague
43. Walker DHT, Davis PR, Stevenson A (2017) Coping with uncertainty and ambiguity through team collaboration in infrastructure projects. Int J Project Manage 35(2):180–190
44. Waseem D, Biggemann S, Garry T (2018) Value co-creation: the role of actor competence. Ind Mark Manage 70:5–12
45. Wassermann S, Reeg M, Nienhaus K (2015) Current challenges of Germany's energy transition project and competing strategies of challengers and incumbents: The case of direct marketing of electricity from renewable energy sources. Energy Policy 76:66–75
46. Winch G (1989) The construction firm and the construction project: a transaction cost approach. Constr Manag Econo 7(4):331–345
47. Yin RK (2002) Case study research: design and methods, (Applied Social Research Methods, Vol. 5)
48. Zhang L, Cao T, Wang Y (2018) The mediation role of leadership styles in integrated project collaboration: an emotional intelligence perspective. Int J Project Manage 36(2):317–330

Chapter 12
Sustainability for Construction Projects

Hans Knoepfel

Abstract Construction projects normally last some years. Their results, buildings and infrastructures, are used and operated during many decades. Meeting the needs of current and future generations is asked for. To deal with the uncertainty of a long-time future and to convince people of sustainable investments is not easy. In this paper, first a basic model and a conceptual background for construction projects is presented. Then the sustainability evidence is extracted from six large, medium size and small construction projects of the past decades. Observations concerning capture, society, economy, environment and resilience are described for each case. The conclusions contain general criteria for the sustainability of a construction project design and resilience as well as suggestions for further research.

Keywords Sustainability · Feasibility · Construction · Value analysis · Long-term outcomes · Impacts

12.1 Introduction

The buildings and infrastructures are used and operated during many decades. The objective of this paper is to contribute approach and experience for agreeing on defined sustainability requirements and objectives and for achieving them for small, medium size and large construction projects and programmes. Meeting the needs of current and future generations is inquired [2, 10, 11].

The large part of the paper is based on the experience over a long-time span [6, 12].

H. Knoepfel (✉)
Rosenthaler + Partner AG, Zurich, Baumackerstrasse 24, 8050 Zurich, Switzerland
e-mail: kn@rpag.ch

R. Cuevas et al. (eds.), *Research on Project, Programme and Portfolio Management*, Lecture Notes in Management and Industrial Engineering,
https://doi.org/10.1007/978-3-030-60139-3_12

12.2 Research Method

The paper deals with construction projects and programmes. From the literature and experience the basic research model was designed (Fig. 12.1)

The basic model expresses

1. the thinking in terms of systems which produce outputs from inputs.
2. the physical system which is used, operated and maintained.
3. the general structure for the sustainability consisting of the social, economic and environmental aspect.
4. renovation and adaptation projects improve the sustainability of the physical structure.

This model leads to the following three general hypotheses:

Hypothesis 1: The *physical structure*, its usage and the characteristics of the subsystems and components as well as the context conditions are a suitable basis for the investigation of the sustainability.

Hypothesis 2: The *social, economic and environmental aspects* are widely used as coarse structure for the sustainability; the detailed criteria for the sustainability of a design vary for the individual project and programme, but *checklists* can be useful for their identification.

Hypothesis 3: With the *resilience* it can be observed if the physical structure can cope with uncertainty.

The *Multi-Case Study* research methodology is used for the investigation of the sustainability. Observations from *six Cases*, Table 12.1 are described and evaluated.

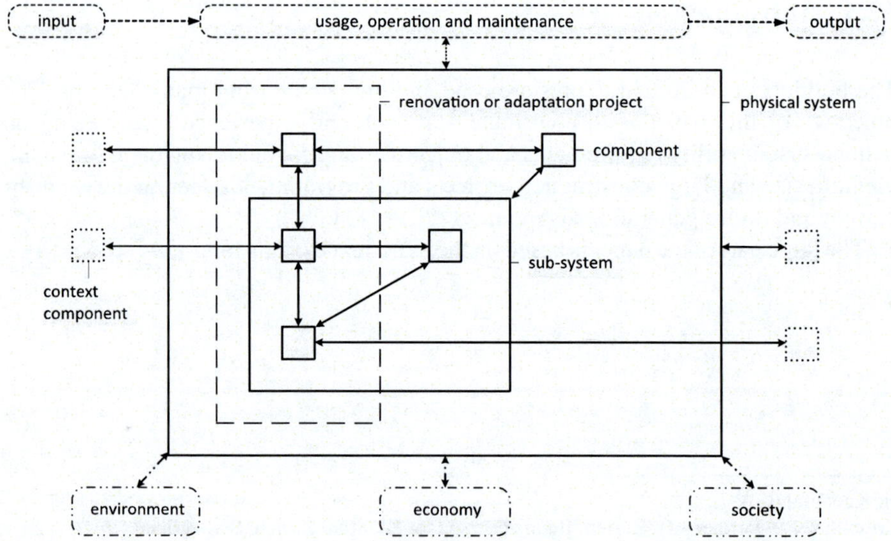

Fig. 12.1 Basic model for the investigation of the sustainability of buildings and infrastructure

Table 12.1 The six Cases

No.	Name	Type	Place	Operation since
1	NEAT/AlpTransit programme	Two New Railway Lines (tunnels 57, 35, 15 km)	Swiss Alps	2008, 2016, 2020
2	Passive House	New Residential Building	Kamakura, Japan	2009
3	Shopping Centre Glatt	New Shopping Mall	Wallisellen, Switzerland	1975
4	Sewage Treatment Plant (City)	Extended Industrial Facility (new generation)	Zurich, Switzerland	1985
5	Whole Sale Market (with accesses)	New Mall (fruits, vegetables)	Zurich, Switzerland	1980
6	Central Laundry	New Industrial Building	Berne, Switzerland	1972

The reason for using this methodology is that the hypotheses are quite general. *Statistical* evidence would require detailed criteria and corresponding samples of projects and programmes. With an investigation using the *Single Case* methodology more depth would have been possible. However the information for several cases was available and the goal of this investigation is to find an approach for defining and agreeing on the whole sustainability requirements and objectives for construction projects and programmes, using own experience of the author from different kinds of real construction projects and programmes. The paper evaluates experience from the past 50 years.

12.3 Conceptual Background

The conceptual background for influencing the sustainability can be presented in two parts

1. *Capture* displays the physical system and its usage, operation and maintenance
2. *Foresee, organize, design and permit* display the way how the sustainability is designed and agreed in a construction project or programme

After describing the capture, general criteria for the sustainability of the physical structure of the whole project design will be elaborated. The investigation of process steps is left to further research; a general description of the process is delivered here as background for this chapter.

12.3.1 Capture

A key question for a construction project is "what is it"? The response is the *content* structure. This structure is outlined in the first phases following the analysis of the existing unsatisfying state and the desired future state, based on the conceptual draft of the usage and needs. With the course of the project, the structure becomes more and more detailed and complicated. The new tool used for depicting it is the *Building Information Modelling* (BIM). It produces the drawings for the deliverables planned for the project.

For the sustainability, the structure should be designed with *subsystems* and *components* that can be *separated* for the procurement, production, erection, maintenance, removal, replacing and recycling due to their different product life duration. This severability and separation can be challenged by aesthetics and cleaning requirements. The structure and building model is used during the whole life cycle of the building or infrastructure. Components and subsystems can be refurbished, replaced, removed or added.

The structure of the content is also useful for the management of the time schedule, the investment cost and the opportunities and risks during the project phases.

12.3.2 Foresee

The investor, the users and the operators of buildings and infrastructures want to see a favourable *business* case for the project. The complete success depends on the results during a long-time future. But the context conditions such as the markets, technologies and weather are uncertain. Also the usage, operation and maintenance are an estimate. The tools for coping with the uncertainty are scenario designs and the analysis of opportunities and risks as well as alternative options for the future.

For the sustainability, the *lifetime* for the buildings and infrastructures and their subsystems and components is estimated. The real end of the lifetime happens when either the fitness for use (including fashion, for example) or the economic performance is no longer satisfactory, or if the legal foundation no longer exists. Living with uncertainty and some unknowns is a part of the construction project business.

During the project cycle, an optimum lifetime shall be estimated and agreed with the investor. Also the *recycling* of all components shall be planned to a reasonable level of detail. A future-perfect strategy, working backwards from or forwards to the future is the project goal.

12.3.3 Organize

Typical roles take care of the tasks and responsibilities during the construction project cycle and the usage, maintenance and changing of buildings and infrastructures. Their obligations and deliverables are normally defined in contracts or role descriptions of their organization. The tool for defining the tasks, authority and obligations is the responsibility matrix.

The interest and responsibility for the sustainability is with many *stakeholders* and *actors*:

1. The project investor and owner wants to realize the benefits of the investment.
2. The users want to do their business successfully using the building or infrastructure.
3. The operators want to offer to the users a reliable context at moderate cost and with little disruption.
4. The designers want to ensure a sufficient sustainability for their customers with a limited effort.
6. The sustainability advisers want the check the projects and operations for ensuring an optimal and complete achievement of the sustainability objectives.
7. The permission authorities want that the projects and operations comply to the laws and public regulations concerning the sustainability.
8. The contractors want to fulfil their contractual obligations concerning the sustainability of the construction processes and deliverables.
9. The public wants that the laws and governmental decisions fulfil the valid sustainability requirements.

Quite often the tasks and responsibilities are not well enough documented and controlled. The interest in a long-term future tends to be uncertain and weak.

12.3.4 Design

A good design of buildings and infrastructures requires experience, creativity and evaluation criteria and measures. It is typical for projects that the design is done in several cycles, from conceptual to detailed design. The design is unique to a considerable extent and there are limited opportunities for continuous improvement through repetition. The tools for the evaluation and achievement of an optimum design are checklists and the value analysis based on the project objectives and context conditions.

Thinking beyond the basic triangle of deliverables, time and cost during the project cycle is necessary for a resilient sustainability in the life cycle of the buildings and infrastructures, Silva (2016). During the past decades a big effort was made to define detailed *criteria* and *measures* for the sustainability. Examples are the *Standards SIA (2017)*: 112/1 for Buildings (52 pages) and *SIA (2016)*: 112/2 for Underground

Engineering and Infrastructures (104 pages) of the Swiss Association of Architects and Engineers, as well as *IPB & KBOB (2010)*: Brochure (94 pages) and Factsheets (60 pages) for the Sustainable Real Estate Management of the main Swiss Investors. Both standards use the dimensions society, economy and environment as the capital structure for the sustainability.

12.3.5 Permit

Construction projects need a permit that shall ensure that the building or infrastructure proposals comply with the laws and public regulations. The number of individuals and organizations who can oppose, formally by objection or informally by political pressure, to a building or infrastructure differs by the kind and context of the project. The tools for checking are the list of laws and public regulations and the checklists with the criteria for compliance, for each law and public regulation.

The *responsibility* for the sustainability can be mainly with

- The individual who is in the case of the construction project mainly the investor and his designer.
- The organizations involved in a construction project if they want to comply with their corporate social responsibility.
- The society as the context for every construction project and building or infrastructure operation.

Depending on the focus, the laws and *public* regulations on one hand, or the *private* voluntary standards on the other hand, have the bigger weight.

12.4 Main Findings

12.4.1 NEAT/AlpTransit

12.4.1.1 Capture

AlpTransit, also called NEAT, a mega-size-programme for crossing the Alps by railway for passengers and freight, see Fig. 12.2, is built in the years 1999 to 2020. The West Line (Loetschberg) contains one basis tunnel of about 35 km (in operation since 2008). The Central Line (Gotthard) contains one basis tunnel of about 57 km (in operation since 2016) and one basis tunnel of about 15 km (going into operation in 2020). The tunnels were realized in several parts, with accesses from the side or the top, for shortening the execution time. The programme was divided into more than 400 projects and requires an investment of about 22'000 Mio. CHF. Typical projects and subsystems were the realization sections of the tunnels with

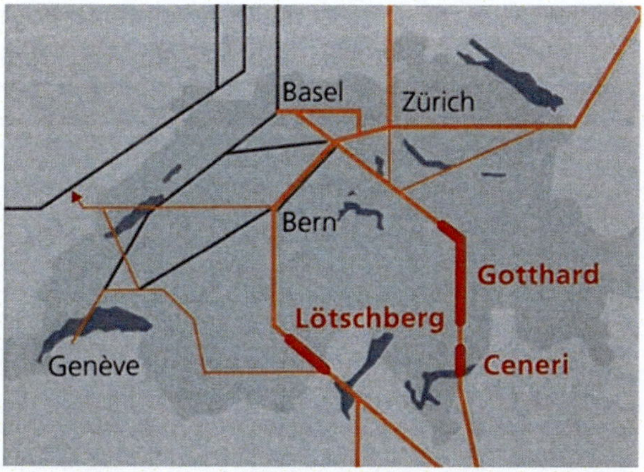

Fig. 12.2 NEAT/AlpTransit, FOT Berne

the tunnel accesses and the excavations, the fills of the excavated material, the engineering, surveying, logistics and supervising subsystems and phases, the subsystems of the carriageways and the railway technology, the supply, maintenance and rescue installations. The main projects were managed by new temporary limited companies owned by the railway companies. Several access line projects were managed by the railway companies themselves. The Federal Office of Transportation supervised and controlled the programme and reported to the Ministry. The Ministry reported to the Parliament.

Input: The rapid passenger trains and the slower freight trains arrive from Northern Europe or from Italy in the South. *Operations*: About 150 to 170 trains per workday drive in a separate tunnel for each direction underneath the Alps with an over-cover up to 2000 m and a temperature up to 35–40 degrees. They are connected with short accesses to the emergency rescue tunnel. After the start of operation of the Gotthard basic tunnel 100’000 trains crossed the Alps in less than 3 years. *Output*: The trains depart on the other side of the Alps.

12.4.1.2 Society

- The Alpine Region is a precious resource in Switzerland for the empowerment, recreation and relaxation of people. A most relevant reason for the programme was to protect this region from the freight truck transportation impact.
- Switzerland agreed to ensure a high-performing transportation through the Alps with the European Union.
- The durable acceptance by the Swiss government and voters was expressed by several successful federal votes for the very large and long-term investment.
- Another positive signal is that the new lines have already been used quite often near their capacity. One day right at the start of operations the passenger train was more than full. The transportation capacity was increased.

- The weight of the safety was higher than the weight of the cost goals; a separate tunnel for each direction rather than one common tunnel was chosen for increasing the safety of the people.
- The passengers appreciate the saving of about 30 to 60 min per trip.

12.4.1.3 Economy

- The two lines satisfy a big part of the historic demand for a secure and rapid trade and traffic, as a part of the European transportation network and in Switzerland. The precedent railway system existed since more than 100 years and was at the end of its technological and economic life.
- The precedent transportation system used loop tunnels and climbed and descended a difference of height of about 500 m for the Gotthard line, for example. This was the reason for high operation costs and long travel times.
- The reliability is increased by avoiding places where avalanches, rivers and landslides and rock-slides can endanger and block the traffic. In addition, with the two main lines rather than one main line, the impact of such events is reduced.
- The places near both railway lines get a significant increase in attraction for businesses and living; opportunities are created for several places.
- With the offer of the new railway infrastructure charging the highway traffic with realistic cost compensation became feasible.

12.4.1.4 Environment

- Changing the model split to less road traffic and more rail traffic is the key long-term goal of the programme, for environmental reasons.
- Eliminating the difference of height (e.g. by 500 m with the Gotthard line) reduces the energy consumption and the CO_2 production as well as noise emission.
- Fillings of excavation material can produce a positive footprint: A new lovely area was built at the shore of a lake, for example.
- Experience from another programme showed the risk, that a dam or wall of a water reservoir could be damaged by the settlement due to the excavation of a new underneath tunnel, also including the risk of a downside disaster due to the flood. The mitigating action was precise geodetic measurement during construction.

12.4.1.5 Resilience

- The rolling stock was a relevant subsystem for the operations. In one of the first days of operation a train was stuck in the tunnel. The tunnel was new and worked perfectly, but another part of the operations system, a not new locomotive did not work in this environment and the train had to be pulled out of the tunnel.
- During the design phases the new European Train Control System was not yet sufficiently developed. The programme management and controlling identified

which subsystems and components depended on it and how the risk on the investment cost and timing should be estimated and controlled.
- The investor required that the project managers included the objectives of the operations, maintenance and further development in the project design. This requirement included the reliability and resilience aspects.
- The future train time tables need to be known for determining some resilience requirements to subsystems.
- A strict design to cost did not make sense for this programme. However the East-West lines of the Swiss transportation system and the traffic near big cities had a high relevance on investment decisions in Switzerland as well. A cost overrun in the programme would create an unwanted shortage of funds for the other places.
- The Federation decided in an early stage to postpone some investment on both main lines of the programme. It is discussed today if the kind of reduction for the West Line was optimal for the long time.
- There was a proposal to add a new project to the Central Line for providing access to a touristic region via the access shaft for the construction. The proposal was rejected based on the value analysis.

12.4.2 Passive House Kamakura

12.4.2.1 Capture

The concept of the passive house has been standardized in the year 1996 by the Passive House Institute in Darmstadt, Germany. The first step was to reduce the effort for heating. The complete goal was a building that is autonomous for heating, cooling, ventilation and electric power. This concept changed many construction subsystems and components. The Japanese architect Miwa Mori went as university scholar to Stuttgart in 1999 and continued studying and working in Germany and Ireland with her focus on sustainable buildings. In the year 2009 she had founded a design firm, realized the first passive house in Japan, Fig. 12.3, and published a book about the standard applied.

Input: Solar energy, temperature differences from subsoil and water of the surroundings. *Operation*: Heating, cooling and ventilation systems, generation of electric power and ecological and effective insulation of the building. *Output*: Recycling of replaced components in the future, maybe heated air and water as well as changed subsoil temperature.

12.4.2.2 Society

- People want to live in a way that is compliant with their sustainability principles.
- The normal life cycle of houses is different in different economies (are houses built in the year 1920, 1940, 1960 or 1980 replaced in the year 2020?).

Fig. 12.3 Passive House Kamakura (*Source* KEY ARCHITECTS Co., Ltd, reproduced by permission)

- People are sensitive to impacts on their health (used air, moisture, radiation, abrasion, etc.).
- Harmful materials and products used in housing construction can reduce the reputation of the owner.

12.4.2.3 Economy

- The price of sustainability can often be reduced by mass selling of new innovative material and building components (standardization).
- Sustainability has to be related to the price, the price for dwellings is often limited.
- The long-time market price for energy is uncertain.
- Many technical solutions for achieving sustainability had to go through several innovation and optimization cycles and seem still not to be mature enough.

12.4.2.4 Environment

- The analysis of many materials and products used in housing construction and the concepts for recycling them is not developed enough.

- The data for the life cycle impact of houses to the global warming should be improved.
- The consumption of external and home-made energy for houses is very different.
- Gardens and plants can fit the local nature and contribute to the diversity.

12.4.2.5 Resilience

- The change of the (family) usage for new and renovated houses during the next 4 to 8 decades should be considered before starting the design phase.
- The subsystems and components of houses (e.g. structural subsystems, facades and roofs, water, HVAC and electrical subsystems, kitchen, baths, finishes, furniture, equipment) have different life times and should be separated for replacement.
- The lifetime for the subsystems and components should be estimated (e.g. fashion cycles: After how many years do floors and furniture in which rooms look too old-fashioned? After how many years are water installations or HVAC or electrical subsystems expected to be replaced?).
- For a renovation project, dealing with corrosion, cleaning and painting are actions with environmental impact.
- Additions to houses are relatively rare (exceptions: stages of investment in case of very high inflation, winter gardens, etc.).

12.4.3 Shopping Centre Glatt

12.4.3.1 Capture

Glatt near Zurich is the shopping centre with the biggest turnover in Switzerland, Fig. 12.4. It opened in the year 1975 and was redesigned in the years 1994–95 without interruption of the operation and achieving even a small increase of turnover during the construction time. With the redesign of the public areas all interior components were replaced by a modern design and the centre was opened for daylight. The final investment cost of 38.5 Mio. CHF was lower than the cost estimate. The scope of the redesign included the 3 levels of shops and the two parking spaces as well as the area for the bus-lines and the change of the adjacent road. Typical subsystems and components were a few changes of the structural system, the adaptation of the ventilation (heating and façade replacement came later) and doors, the replacement of all finishing elements (floors, balustrades, wall and column sheeting, ceilings, light and sound installations) in the shopping areas, access ramps and parking areas, 14 entrances to the centre, bus stations, 2 new road circles, roads. For the project management, a structure of space units, technical elements types and phases was used.

Fig. 12.4 Shopping Centre Glatt, Glatt Wallisellen

Input: The customers arrive by car or bus through 14 entrances, goods arrive by truck, sunlight does not enter directly. *Operations*: The customers circulate walking and using ramps, moving stairways and elevators, no noise from the highway is heard, the turnover is about 600 Mio. CHF per year. *Output*: The customers depart by car or bus, waste departs by truck.

12.4.3.2 Society

- Most customers come by car, to about 4500 parking lots, a minority comes by bus or bicycle.
- The customers do not care about construction work, adequate actions for preventing accidents and not bothering the customers were taken.
- Noise from construction was not an issue, but the risk of a fire was taken most seriously and mitigated by a fulltime independent observation of welding works.
- The shopping center attracts people with an excellent mix of products, a good atmosphere, appropriate opening hours, events and restaurants.
- Parking is free of charge which has been objected by environmentalists, but here is some logic for parking free of charge for a location of the center outside the city, adjacent to a highway, large weight and size shopping.

12.4.3.3 Economy

- The benefit for the customers from the investment in different building components has a different relevance (shops first, parking lots last).

- A modest and nice design of the public area had been appreciated for more than 2 decades, but need to be changed in the 1990ies with the renovation project to maintain a difference to competitors. The design of the shops is often changed in shorter cycles.
- The materials and design of the new surfaces as well as the ingenious daylight and artificial light design made people to enjoy working in the new centre.

12.4.3.4 Environment

- The effort for the ventilation and cooling was reduced substantially by an intelligent natural air circulation.
- Some parts of the building were not at the end of their technical lifetime, but they would have hurt the eye of the customers by reducing the beauty of the new design which became a pleasure for visitors, buyers and shop owners. From the functional point of view the old parts were out of date.
- The design for the bus stops outside the building and the announcement of the departures inside the building improved the usability of the public transportation.

12.4.3.5 Resilience

- With the new connection to and stop of the new fast tramway line from the year 2010 the public transportation became more attractive.
- The North access of the centre was adapted to the new large real estate development of the neighbour in the years 2010 to 2014 and improved again in the year 2018.
- The tower on the top of the shopping centre was not in the scope of the redesign; with a new project, the façade and HVAC of the tower was renovated with an innovative environment-friendly technical design.
- The West access of the centre was adapted to a new office building of the neighbour.
- All shopping centres in Switzerland are challenged today by the online shopping and the customer tourism to other countries. The turnover of this centre remained the same during the past years in spite of new competition.

12.4.4 Sewage Treatment Plant Zurich

12.4.4.1 Capture

The sewage treatment plant of Zurich is the largest in Switzerland and its extension and renovation realized in the years 1980–85, Fig. 12.5, was the biggest investment of the city, with 304 Mio. CHF. The structure of the project on a land surface of about 250 times 1000 meters was detailed in 75 subprojects, also due to many stages of construction and change of operations. The operations had to be sure and safe anytime, during 365 days of 24 h. The main parts of the facility were the renovation

Fig. 12.5 Sewage Treatment Plant Zurich, ERZ Zurich

of the intake, the mechanical treatment and the sludge treatment, the new biological and chemical treatment, the new aqueduct for connections and the new garage and office building. Typical subsystems and components were the construction pits in the groundwater, the structural systems, the piping, the many installations for the treatment and for moving water and sludge, the system for air and gas, the system for the electric power, the system for central and local monitoring and controlling, the roads and sewerage system for the local area.

Input: The sewage arrives in a large canal after collection in the sewer system of the city. *Operations*: Solid objects are picked out and cart off. Wastewater is treated in several steps, sludge also. Heat and electricity was produced with gas. Used and polluted air was treated. *Output*: Sludge was disposed to agricultural fields by trucks. Cleaned water flows into the receiving river. Smell leaves open basins. The maximum sewage input of about 9 m3/sec including rain stormwater as well as the volume of sludge was estimated.

12.4.4.2 Society

- The quantity and quality of the incoming sewage (including the rain stormwater) varied and was hard to control.
- The open tanks could be smelled in the neighbourhood depending on the wind situation. It was discussed if the tanks had to be covered, but this requirement was not made in general by the public authorities.
- The used air from the congestion tanks was burnt with the gas in the gas turbines.

- Inhabitants of the nearby district of the city felt uncomfortable about their view to the new aqueduct, but the public authorities decided that this view to an industrial zone was acceptable and the cost of underground connections cannot be justified.
- The requirements to the water flowing to the Limmat River (receiving watercourse) was regulated by the public authorities with four main parameters.

12.4.4.3 Economy

- The estimate for the incoming sewage quantity was relatively high due to the not yet realized separation and seepage of rainwater, the risk of rain stormwater and the leakage of ground and surface water to the canal system.
- The economic viability depends on the charge on the clean water consumption of the customers. Part of it is the income for the collection and treatment of the sewage. This amount can be adapted only on medium term if at all.
- The public limited company had to borrow money from the city for a large part of the investment.
- The owner and operator insisted on a satisfying long-time performance of the sewage treatment plant. If no agreement within the project was possible, the issue had to be escalated to him and maybe to the City Authorities supervising the public limited company.

12.4.4.4 Environment

- Critical requirements to the water quality in the river were bathing of people and living of fishes. The State of Zurich had to establish the requirements to the outflow from sewage treatment plants.
- An incident during construction work causing the flow of sewage into the river was a capital project risk.

12.4.4.5 Resilience

- The rain stormwater inflow was a reason for the mayor risk of washing out the biological and chemical treatment tanks and the pollution of the receiving watercourse. A new large tank was realized under the island in the river near the extended sewage treatment plan in the years 1984–88 with an investment of 48 Mio. CHF. With it, the additional sewage from the North of Zurich could be treated (see next bullet).
- A proposal for integrating the old treatment plant for the North districts of the city of Zurich into a new regional plant was not accepted by the voters of a village involved. The analysis led to the conclusion that the sewage could be conveyed to the large treatment plant extended already, with a tunnel. This project was realized in the years 1994 to 2001.

- Additional requirements (in total 23 parameters) to the outflow led to more treatments at the end of the plant. With ozoning remaining medicines etc. were removed and with filtration very small suspended matters were withheld.
- The gas turbines turned out to be uneconomical; they were no longer used after some years.
- The usage of the sludge in agricultural areas was forbidden by the Federal Government from begin of 2003. Already from 1985 dumping of sludge in a landfill site had been stopped. The sludge had to be burned at several places outside Zurich. It was decided to burn the sludge of the whole Kanton of Zurich in the sewage treatment plant of Zurich City. The investment was about 65 Mio. CHF and the operation started in 2015, with a maximum capacity of 90'000 tons per year.
- In the future the phosphorus should be recycled from the ashes.

12.4.5 Zurich Engros Market

12.4.5.1 Capture

The Fruit and Vegetable Market in Zurich-West was built from mid 1977 to mid 1980, Fig. 12.6, for delivering fresh fruits and vegetables to re-sellers and large users, in the early morning hours. The investment including additions for the operations was 48 Mio. CHF. Earlier the area was used as clay mining and industrial waste disposal pit. The project was accepted based on a realistic feasibility study and the urgency of removing the selling from railway wagons due to the extension of the central railway station of Zurich. A legal unit each for the investors and operations was founded. Main

Fig. 12.6 Zurich Engros Market, ZEMAG Zurich

buildings were a subsoil level for unloading and storing fruits and vegetables and a ground floor level for the market and the loading of the customer vehicles and the unloading of the railway wagons as well as a ground-level local producers market and an entry, office and restaurant building. Typical subsystems and components were the reinforced concrete structure, façades and roofs, HVAC, elevators and stairs, interior separation walls and finishes, shelfs, electric transportation vehicles, railway access system, platforms, roofs, roads, parking areas, fences.

Input: The fruits and vegetables arrive in the late evening by train or truck. The customers arrive in the very early morning with their private vehicles. *Operations*: The products are stored, sorted, presented and moved mainly in the building, transported with electrical vehicles and elevators. The turnover per day is up to 800 tons. *Output*: The fruits and vegetables depart in the early morning. There is light and noise during the market time. The empty trucks leave later.

12.4.5.2 Society

- The merchants accepted the new attractive central market including risk and opportunity of the increased rate of competition.
- The inconvenience due to the early morning noise from the market was eliminated with the localization outside the city centre.
- The local producers appreciated that they got a roof for placing their vehicles and presenting and unloading their products.
- The project team and context accepted the reasons for appointing a civil engineer (rather than the architect) as project manager due to the nature of the project.

12.4.5.3 Economy

- The project logistic planning assumed that about 50% of the incoming fruits and vegetables from the Swiss and Southern European producers would arrive by railway and the other 50% would arrive with large trucks. All customers would use their small or large private vehicles.
- The project manager helped the investors and merchants to establish a medium-term business plan and requirements and ideas for financing the project as well as achieving improvements for the organization, operation and viability of the market.

12.4.5.4 Environment

- The contaminated site and the groundwater flowing below it required a careful management of the excavation for the building pit.

- For the construction permission it was crucial to design the connection to the main street via an access street (not directly which would have disrupted the morning traffic with an additional traffic light).

12.4.5.5 Resilience

- The reality during operation was that the incoming fruits and vegetables arrived by trucks. The railway connection was no longer used at all from the year 1990.
- The banana came in not mature enough for selling. A banana ripening installation was added in the subsoil level.
- A ramp to the roof level as well as the structural design for rooms on this level had been foreseen. Some rooms were added on the roof. Later the ramp was combined with the neighbours new access which became necessary due to an extension of the main street.
- The inside of several rooms is no longer looking as fresh as the fruits and vegetables: A renovation was undertaken recently. This renovation is also extended to the façade and windows saving energy and reducing noise inconvenience.
- There is still a discussion about the additional usage potential of the relatively large and only temporarily used area. The first idea of an additional several level building on the area was not realized.

12.4.6 Central Laundry Berne

12.4.6.1 Capture

The laundry for hospitals, retirement homes, etc. in the region of Berne was built in 1970–72, Fig. 12.7, with an investment of 21.5 Mio. CHF. It followed a trend in Central Europe for the centralization of this business rather than doing it in each hospital, retirement home, etc. The project was accepted based on the feasibility study of an architect and designer. For the organization of the investment and operations, a public owned limited company was used as an innovative format. The building location was an industrial area not far away from a main hospital and the city's waste burning plant. The building was composed of a ground floor for intake, sorting and checking the used laundry and a first floor for the production lines of cleaning, pressing, repairing, folding and packing. Typical subsystems and components of the plant were the structural system bearing the heavy machines, the facades and shed roof, doors, platforms, several large and small equipment, movers, HVAC, interior separations and finishes, transportation vehicles and trucks, access road, parking lots and surroundings.

Input: The used laundry arrives from the customers by truck. *Operations*: In the building it is stored, sorted and moved with sacks and conveyors. The capacity per day was about 20 tons (in 8 h). The super-heated steam comes in by tubes from the

Fig. 12.7 Central Laundry Berne, InoTex Berne

waste burning plant. *Output*: The clean laundry is stored in shelves and departs by cleaned trucks. The warm waste water flows into the canalization of the city.

12.4.6.2 Society

- The hospitals and homes etc. asked for avoiding the risk of recycling viruses and other sources of diseases during the collection and storage of the used textiles or even washed textiles was at least as low as the risk in their in-house laundries.
- The parliament accepted that the organization as a limited company with public shareholders was economically and socially feasible and efficient.

12.4.6.3 Economy

- The central laundry could save money with the industrial process and high occupation, in spite of the additional transportation costs.
- With the location in an industrial zone the production could be laid out on two levels and working with daylight, due to the lower cost of the land. Once loaded and unloaded, the additional cost due to a larger distance is low.
- Reflecting on the choice of the location, an even more remote building with one level only may have been a good option, if the distance of the work place and the heat transportation cost and losses would have been feasible.

12.4.6.4 Environment

- The central laundry used space in an industrial zone while a hospital laundry used spaces in a more expensive down town areas.
- The building was placed not far from the City's waste burning plant that delivered super-heated steam for the wash process. This was an environment-friendly usage of available energy.

12.4.6.5 Resilience

- The central laundry expanded their business in a first step by cleaning professional clothes. In a second step of integration they completed their service to the customers by producing and selling innovative professional textile clothes. The organization was renamed from "Laundry" to "InoTex".
- In the year 2013 the organization was integrated in an international group with head quarter in France.
- In the year 2007 handicapped-accessible workplaces were achieved. For improving the ecological performance, three projects were realized: Heat recycling in the laundry, photovoltaic installations, replacing energy from gas to energy from steam.

12.5 Conclusions

12.5.1 Hypothesis 1

It was useful to describe the physical structure for each case in the appropriate detail as basis for the investigation of the project's or programme's sustainability.

The basic model can depict the system structure.

The criteria for the sustainability (such as the expected lifetime, the acceptance by the users, the suitability for replacements and the lifetime costs) are defined as the characteristics of the subsystems or components.

The main uncertainties of the fitness for usage stem from the uncertainties of the future usage and context conditions.

12.5.2 Hypothesis 2

For the general criteria for the sustainability of a construction project design, the following checklist can be condensed from the cases:

(1) The sustainability is interacting with strategic business decisions such as centralization vs. decentralization and conceptual design of the services, manufacturing and distribution.
(2) Construction projects can improve the environmental sustainability by reducing energy consumption, CO_2 production, noise, floods, etc.
(3) Construction projects can be economically sustainable by providing an appropriate infrastructure with economic life cycle cost.
(4) Construction projects can achieve the social sustainability by the acceptance of their output and their beneficial outcomes and impacts.
(5) The choice of the location of buildings is a general opportunity for improving the sustainability.
(6) The layout and design of a building has an impact on the competitiveness of its users.
(7) A key success factor of construction projects in the innovation and sustainability analysis of materials, components and subsystems.
(8) Infrastructure projects are often related to the sustainability of the transportation of goods and passengers.
(9) Infrastructure projects are often related to regional sustainability aspects.
(10) Social sustainability can be demonstrated by the analysis of the real usage and acceptance by the stakeholders.
(11) The sustainability of a current project can depend on the contribution of subsequent projects.
(12) The estimates for the future business performance and the future changes have an impact on the economic sustainability of a construction project.

12.5.3 Hypothesis 3

For the general criteria for the resilience of the output of a construction project concerning its sustainability, the following checklist can be condensed from the cases:

(1) The sustainability depends on the future development of the technology, ecology and society; this will challenge the resilience of the project output.
(2) The sustainability depends on the future, uncertain usage of the project output; this will challenge the resilience of the project output.
(3) The critical stakeholders for the resilience are the owners and financing partners, the users and the operators of the project result.
(4) Potential or planned subsequent projects connected can contribute to the resilience, in connection with the output of the current project.
(5) The reliability and resilience of a system depends on all new and existing components and subsystems for the usage of the system.

(6) The resilience is a key success factor for the project selection and the investment decisions.
(7) Documented requirements and measurements facilitate the continuing sustainability analysis and management.

12.5.4 Further Research

The *structure and criteria for the sustainability for construction projects* should

- be discussed and improved (e.g. comparison with [2]).
- be combined with International Competence Standards for project, programme and portfolio management (e.g. [3]).
- be anchored in the project, programme and portfolio organization (e.g. responsibility for the sustainability).
- deal with the uncertainty of the long-time future of the construction deliverables (see, e.g. [9]).
- deal with the different life cycle durations of subsystems and components (see, e.g. [1], [5].
- be aligned to the project life cycle (e.g. its processes, activities, documents).
- be detailed with checklists (e.g. [4, 7, 8]) helping to ensure a sufficiently complete and realistic sustainability of the design proposals and permits.

The following *methodological approaches* could fit

- compare existing approaches, models and baselines (theoretical research).
- investigate selected general issues (such as responsibility for sustainability, different life cycle durations, uncertainty) with further multi-case studies.
- get detailed checking through single case studies.
- establish quantitative statistical evidence for suitable criteria (e.g. the footprint of construction materials).

Sustainability will continue to be an interesting field for research.

References

1. Driver P, Seath I (2015) There are no shortcuts from Projects to Benefits. PM World J
2. GPM Global (2019) The P5 Standard for Sustainability in Project Management. Version 2.0. Green Project Management
3. IPMA (2015) Individual Competence Baseline for Project. Programme and Portfolio Management. Version 4.0 (IPMA ICB4)
4. IPB & KBOB (2010) Nachhaltiges Immobilienmanagement. BBL Berne
5. Knoepfel H (2010) Sustainability and competence elements. IPMA Expert Seminar. Zurich, 20–28
6. Martinez-Almela J (2010) Agribusiness: primary production, retail and distribution. IPMA Expert Seminar. Zurich, 187–222

7. SIA (2016) Nachhaltiges Bauen—Tiefbau und Infrastrukturen. Swiss Standard, Zurich
8. SIA (2017) Nachhaltiges Bauen—Hochbau. Swiss Standard, Zurich
9. Silva M (2016) Thinking outside the triangle. IPMA Expert Seminar. Zurich, 247–260
10. Silvius G, van den Brink J, Köhler A (2010) The concept of sustainability and its application to project management. IPMA Expert Seminar. Zurich, 131–145
11. Tam G (2010) Sustainability competence requirements for project manager. IPMA Expert Seminar. Zurich, 175–185
12. Xue Y, Knoepfel H (2008) Bring values to the public. IPMA Expert Seminar. Zurich, 44–60

Chapter 13
Segway Model for the Assessment of Megaproject Excellence: Project Excellence Baseline Approach to the Korea's Pilot Smart City Projects (Sejong 5-1 Life Zone and Busan Eco Delta Smart City)

Jaehyun Lee, Unho Lee, Eunsang Yoon, and Changwoo Park

Abstract The fourth industrial revolution and technological advances have been accumulated into the development of smart cities, which has been influencing the shape and value of human life. South Korea has implemented the national project of building a model smart city that is futuristic and innovative. Two megaprojects are underway the Sejong Administrative City (Sejong 5-1 Life Zone) and the Busan Eco Delta City (Busan Eco Delta Smart City) and have been designated as the national model cities. This study aims to develop a model for assessing the excellence of megaprojects. For this purpose, we analysed the case of smart city cases of National pilot smart city. International Project Management Association (IPMA) has provided Project Excellence Model (PEM) for assessment of project excellence. However, megaprojects like smart cities have a completely different character compared to general projects that have existed. It is necessary to develop a new model that is appropriate for assessment of megaprojects. We propose the Segway model to evaluate megaprojects. The Segway model can provide a meaningful contribution for

J. Lee · C. Park (✉)
Department of Applied Engineering, Graduate School of Engineering Practice, Seoul National University, Building no. 38, 1, Gwanak-ro, Gwanak-gu, Seoul 08826, Republic of Korea
e-mail: cwpak@snu.ac.kr

J. Lee
e-mail: homosapiens@snu.ac.kr

U. Lee
Institute of Engineering Research, Seoul National University, Building no. 38, 1, Gwanak-ro, Gwanak-gu, Seoul 08826, Republic of Korea
e-mail: egypt-lee@snu.ac.kr

E. Yoon
Program Management Consulting Company of Sejong Administrative City Construction (HKCMC), Government Complex Sejong, 94 Dasom 2-ro, Sejong-si, Republic of Korea
e-mail: esyoon123@gmail.com

R. Cuevas et al. (eds.), *Research on Project, Programme and Portfolio Management*, Lecture Notes in Management and Industrial Engineering,
https://doi.org/10.1007/978-3-030-60139-3_13

“the assessment“ of megaproject excellence by analysing the connectivity of the megaproject success factors and PEM.

Keywords Project excellence · Megaproject · Smart city · Segway model · Case study · Korea

13.1 Introduction

The recent trend of urbanization is to make a city “smart”. So-called “smart city” has been on the rise in an attempt to address various issues that most cities are confronted with congestion, inconvenience, pollution, etc. These issues might be better tackled by the help of city’s functions combined with new technologies such as big data application, information communication technologies (ICT) and interconnectedness. A smart city involves diverse technologies and stakeholders. Naturally, planning and building a smart city becomes more massive and complex. This brings about changes on the structure of preexisting urban planning, having an impact on the way we manage a smart city project.

Given its scale and complexity involved, smart city has the characteristics of megaprojects. Megaprojects have been much studied by renowned institutions such as IPMA, Project Management Institute (PMI), Oxford University, etc., but it remains as one of the difficult concepts that can clearly be defined in the field of project management. This is mainly due to the fact that a former definition no longer accommodate the qualities of newly emerging megaprojects combined with development of new environment and technologies. A smart city project is a good example preexisting understanding about megaprojects may not be suitable for it. Considering that various approaches and examples should be further studied to properly establish the study of megaprojects, research on the project management of smart city cases is expected to make an important contribution to the development of megaprojects in the twenty-first century.

This study aims to propose how to manage the increasing complexity of megaprojects using two ongoing smart city projects in South Korea. In 2019, the Korean government designated (i) the 5-1 Life Zone of Sejong City (the new administrative city of Korea) and (ii) the Eco Delta Smart City in Busan as smart cities to engender momentum for national growth currently, specific urban planning and policymaking for these smart cities is currently underway. These two cases will show the importance of project management and the difficulty of managing organizational complexity. Finally, this research proposes a model that takes into consideration the application of smart city technologies for the operation of IPMA Project Excellence Baseline (PEB) Standard to properly evaluate and manage smart city projects.

13.2 Literature Review

13.2.1 Evaluation of Project Performance

Theoretical and empirical researches on project have continually developed various factors for promoting successful project performance [4, 18]. The findings from those studies have been employed to build a framework for successful project execution, and the framework has been improved continuously based on new project research [4, 18].

IPMA PEM was developed in 1996 by the German Project Management Association (GPM). The PEM was based on the Total Quality Management (TQM) and the European Foundation of Quality Management (EFQM)-Excellence Models [8, 21]. However, the EFQM-Excellence Model, developed in 1989, was difficult to be applied to all situations due to the unique characteristics of each and every project [23].

Bou-Llusar et al. [1] proposed an improved model by embedding social and technical dimensions of the TQM framework into the EFQM-Excellence Model. Farahani, Sobhieh and Khameneh [5] identified the interaction by mapping elements of the ICB onto the domain of PEM and conducted an exploratory research on the points of reforming the PEM by analysing individual competencies, project integration and their relationship. Westerveld [23] proposed five types of Excellence Models according to the project type, but it would be necessary to develop additional models to cover new, various characteristics of projects. IPMA announced an improved model of PEM in 2016 to reflect new trends to be considered for a project such as sustainability and environment, creating IPMA Project Excellence Baseline (PEB) [8].

Park [17] proposed a conceptual framework for the Project Ecosystem Competency Model to promote sustainable performance in engineering organizations and industries while enhancing the value of project stakeholders for project sustainability.

With new characteristics and trends being established for determining project success, research on project performance must continue. This is particularly true with a sizeable, complicated project like in the case of megaproject, however, research on megaprojects is still insufficient. It is necessary to study and identify the success factors for megaprojects in order to propose how to improve the PEM accordingly.

13.2.2 Megaprojects

Megaprojects do not stand for merely large projects. Its large scale is related to various elements such as cost ($1 billion or more), time of construction (many years to build), stakeholders (multiple and diverse), transformation (changes as the project develops), and influence (on the lives of a large number of people). The specific scope of each element varies according to the scale and type of megaprojects. Defining whether it constitutes megaprojects need some consideration of a number of elements relevant to the qualities of a particular project, since no criteria are suggested as clearly defining what exactly megaprojects are. Even with regard to cost, Oxford suggests that megaprojects are a billion dollars or more [6], but the European Union defines it as exceeding € 200 million [9]. The definition of megaproject changes by time as well the construction of a 50-floor building might have been regarded as a megaproject 100 years ago, but it is not the case any longer. Another defining element of megaprojects that may easily be overlooked but important is complexity. Massiveness alone does not represent megaprojects, since a huge but simple project can be just referred to as a large project. Nonetheless, complexity can accumulate by the increasing scale, scope, time or/and participants of a project. For example, the participation of multiple stakeholders makes a project difficult to be governed. Similarly, a great number of interdependent non-autonomous subprojects contribute the overall project complexity and increases coordination costs. Levitt and Scott [14] further add the interdependency between subprojects as contributing to coordination complexity, that is, a high level of interdependence requires intensive mutual adjustments to coordinate between subprojects. Increased complexity can arise from the project's spatial or technical configuration, lack of the standardized technologies involved in the project, political or economic impacts on a region involving various stakeholders, and multiple participants based on different institutional frameworks, etc. [14]. The complexity of a project can also be affected by the difficulty of the project, outcome variability and non-linearity, and non-governability [13].

Escalation in complexity and scale of a project poses a number of challenges, making the project riskier. The project scope or ambition level can change significantly over the long project cycles of megaprojects. Mismanagement of costs, schedules, benefits and risks once accumulated can result in serious cost overruns, delays and benefit shortfalls, undermining project viability during project delivery and operations [6]. Megaprojects require enormous and constant efforts of coordination among diverse stakeholders and subprojects failing such coordination can cause conflicting interests and delays. This is why effective and factor-oriented project management is especially crucial for megaprojects.

13.2.3 Smart Cities

Smart city is a good, contemporary example of megaprojects considering its scale and complexity. Smart city aims to provide its citizens with optimal city operations and services by means of various types of electronic Internet of Things (IoT) by collecting and using necessary data to manage public assets and resources efficiently. It is the cluster of cutting-edge technologies that covers an extensive range of urban functions from transportation, residences' use of energy resources and residential cost management. A smart city requires a massive amount of effort and investment focused on the installation of ICT and various physical devices/infrastructure connected to an IoT network in order to achieve its intended functions enhancement of quality, performance and interactivity in urban services through real-time responses based on the real-time data provided [11]. This involves a billion-dollar, gigantic-size project, a number of which are underway around the world. For example, India's smart city mission is estimated to require investments worth 150 billion U.S. dollars over the next few years [3]. McKinsey's 2017 Global Infrastructure Initiative Report [15] says the King Abdullah Economic City, in which the use of technology is a major component of developing this $100 billion project, is the largest new city development project to date in Saudi Arabia [7]. The City of Songdo in Korea, which received advanced digital technologies provided by Cisco and was presented as being 100% connected, was developed with a budget of 40 billion dollars. Masadar in Abu Dhabi costs 22 billion dollars. Rio de Janeiro in Brazil was designed by IBM for 14 billion dollars [20].

A smart city project can be more complex than other megaprojects for a number of reasons. Firstly, open networks and citizens' active participation, which are the key features of a smart city, make the decision-making and coordination processes more complex. Effective project management for a smart city will thus depend on how to reduce the time, efforts, and costs of coordinating many participants and stakeholders during each stage of the project construction and operation. Secondly, the need for the interconnectedness of numerous technologies applied to every corner of urban design and life increases the level of interdependency among the subprojects comprising the overall project. The interconnectedness of ICT devices, infrastructures, data and services makes it difficult to partition the project into a number of autonomous subprojects in particular, the lack of standardization or incorporation of advanced technologies into the system architecture can create a large number of unnecessary interdependencies and coordination complexity for a system integrator to manage [14]. Thus, it is important to develop and apply IoT solutions based on interoperability standards to enable various IoT devices from different manufacturers to communicate and share data [2]. Lastly, as smart city technologies bring in technical expert groups as key actors in smart city project management and operation, complexity increases in terms of organization as well as technology. Many have paid attention to new technologies themselves as appearing to challenge a smart city project, but in fact, it is the organizational complexity that poses a real challenge. Organizational complexity has more significant impact on project performance than technological complexity [24], because the former requires mutual and constant coordination for an effective performance of the entire project, while the latter can be managed by technical expert groups related to particular technologies or network. However, there has been a lack of study on the megaproject management approach to smart city development that focuses on the management of organizational complexity.

13.2.4 IPMA Project Excellence Baseline (PEB)

There are a number of management models that help project managers deal with large and complex projects. One of the widely used methods to measure, and to eventually enhance, the performance of a project is the PEB. The IPMA PEB aims to provide a guide for organizations in assessing the success of a project while describing the concept of excellence in managing the project 8]. The specific standard is called Project Excellence Model (PEM). It establishes an analytical system of points for mainly three areas of a project: (1) people and purpose; (2) process and resources and (3) project results. (1) deals with leadership and value, objectives and strategy, and project team, partners and suppliers. (2) focuses on project management processes and resources like time, cost, quality, etc., and management of other technical processes and resources such as engineering design, safety and environment. (3) covers customer satisfaction, project team satisfaction, other stakeholders satisfaction and project results and impact on environment.

The PEM, as a qualified management model, can be used extensively for various kinds of projects [17]. This does not mean, however, that the PEM is designed to properly measure the excellence of megaprojects. As discussed above, megaprojects are distinguished from other projects in terms of its scale, time, cost, complexity, stakeholders, impact, etc. The larger the scale or complexity of a project becomes, the more it is likely to generate an area that cannot be measured. It is thus necessary to partition a huge chunk of the project into decomposable subprojects based on critical areas relevant to the success of a particular project. Instead of applying the PEM to their megaprojects as a whole, project managers can examine people, purpose, process, resources and results of specific but major subprojects or areas of megaprojects.

13.3 Methodology

This research is based on literature review and qualitative case study to improve the preexisting PEM. We examined multiple sources of evidence regarding smart city cases such as conference materials, research papers, policy reports, legislation on smart city and other case studies [25]. Based on the analysis of the literature review and smart city project data, we came to draw the inference that the PEM for a megaproject requires a new approach that takes into account the characteristics of a megaproject.

13.4 Case Study

13.4.1 Smart City Pilot Projects in Korea

Development of smart cities in South Korea dates back to the early 2000s under the name of the Ubiquitous City: "U-City" as shown in Table 13.1.

Table 13.1 Development of smart cities in Korea. (*Source* [12] with author's revision added)

Period	First generation (2005–2010)	Second generation (2011–2016)	Third generation (2017–current)
Name	U-City 1.0	U-City 2.0	Smart city
Main systems	Individual infrastructure + service	Infrastructure-based service	Service infrastructure as an ecosystem
Service innovation	Provider-centred	Provider-centred + some users	User-centred + AI
Data availability	Closed	Closed and partially open	Fully open and partially closed

It was one of the first and leading cases where State's government makes policy support for introduction of ICT and ubiquitous environment to urban planning, which received a global attention [10]. It is reported that more than 50 municipalities have built or carried forward the U-City projects nationwide as of mid-2010 [16]. However, following problems were found in the first and second generations of U-City projects: the projects were mostly for new city construction that requires a high-cost investment; there was a lack of experience of necessary services due to the business led by public institutions; there was no unified driving system at a national level; and the effort was inadequate to coordinate the relationship between the existing U-City regulations and other related regulations.

Since 2017, Korean government has carried forward the national pilot projects of building a futuristic and innovative smart city in the two regions the Sejong Administrative City and the Busan Eco Delta City (Table 13.2). The project of Sejong consists of interactive services for citizens such as autonomous driving and shared mobility, health care and education, along with a spatial plan that is optimized for Seven Innovation Services mobility, health care, education, energy and environment, governance, culture and shopping and job. Busan Eco Delta Smart City focuses on developing proposals on smart water management and on "Five Innovative Clusters" for fostering new industries related to the fourth Industrial Revolution.

13.4.2 Decision-Making Structure for Smart Cities

In a traditional urban planning, decisions are made vertically from the Korea's Ministry of Land, Infrastructure and Transport through municipal governments down to project operators, as shown in the left side of the Fig. 13.1. This makes the communication between actors easier and simpler, but cannot encompass multiple, public and private actors in the decision-making process. A smart city, on the other hand, cannot help but pursue a horizontal decision structure to combine new decision-makers and reflect the needs or interests of various stakeholders including citizens.

Table 13.2 The current state of government-driven pilot smart cities in Korea

Project title	Sejong 5-1 Life Zone	Busan eco delta smart city
Cost/Business period	$1.2 billion/July, 2018–2022	$1.8 billion/July, 2018–2023
Planned area	2.7 km^2	2.2 km^2
Operator	LH (Korea land and housing corporation)	K-water (Korea public water management company)
Vision/Philosophies	City as a sustainable platform to enhance civic happiness and provide creative opportunities	Bring forwards futuristic living where the nature, people and technology come together
Three innovations	– Happy citizens – Sustainable city – Creative opportunities	– Process innovation – Technological innovation – Governance innovation
Key challenges	– Seven innovation services (1) Mobility (2) Health care (3) Education (4) Energy/Environment (5) Governance (6) Culture/Shopping (7) Jobs	– Three focused strategies (1) Smart tech city (2) Smart water city (3) Smart digital city – Seven main contents (1) People-oriented smart city (2) Citizen engagement (3) Living network (4) R&D plug-in city (5) Regulatory sand box (6) Big data and open data city (7) Smart 4IR technology

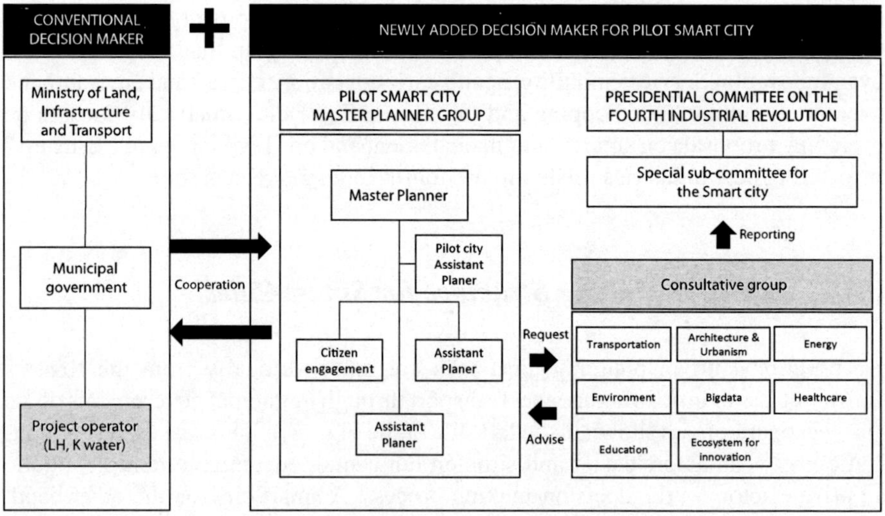

Fig. 13.1 The structure of national pilot projects for smart city. Status of LH smart city and national pilot smart city (LH Corporation), presented at the third Engineering Project Management Congress at Seoul National University, with author's revision added)

Prominent features are (1) the inclusion of the master planner to address the coordination issue observed in U-City projects, and (2) the participation of an external expert group due to the important functions of smart city technologies.

According to the pilot smart city plans, various decision-makers participate in the projects. This is highly likely to bring about some difficulties in communication and decision-making processes. In order to address this issue, it is necessary to have a project manager who can manage and direct various organizations. Considering the importance of coordination among various stakeholders, the authorities should establish a special organization, which can orchestrate various stakeholders, while dispatching competent professionals to engage in decision-making processes and communications with each stakeholder. As more smart city cases emerge, it will become increasingly important to evaluate the effectiveness of various approaches to smart city project management.

13.4.3 Proposal and Conclusion

We propose the "Segway model" to improve the evaluation quality for the excellence of megaprojects. The Segway model, named for its figure of connecting two different areas as illustrated in Fig. 13.2, is intended to make a PEM evaluation on critical subjects of megaprojects by connecting the two for a more proper and accurate evaluation. This enables us to identify and address the sources of poor performance, and ultimately, increase the likelihood of achieving a successful outcome for the project.

The key to the construction of a more effective and useful Segway model lies in the designation of critical subjects relevant to the quality of megaprojects. For example, general subjects for urban planning such as the environment and transportation are

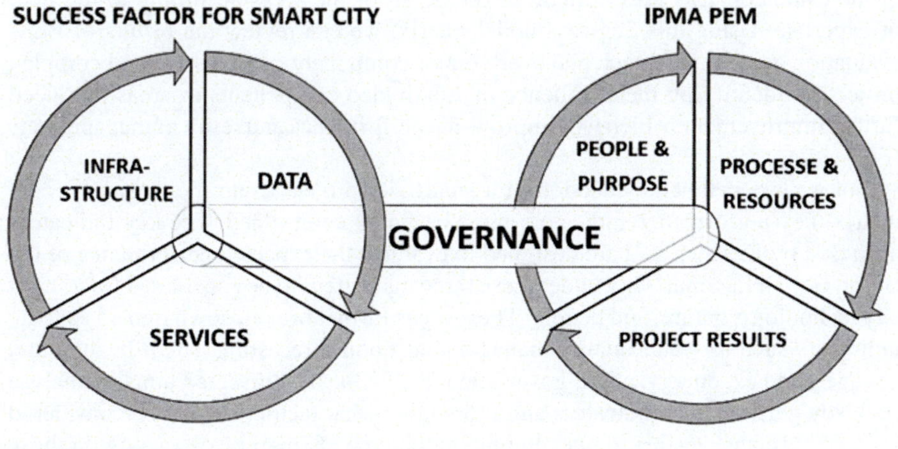

Fig. 13.2 Segway model for megaprojects

neither appropriate nor relevant to the evaluation of a smart city project. A Segway model targeting such general subjects would not make a significant difference as a measurement tool for smart city projects. As seen below, the key components promoting the achievement of smart city objectives and its competitiveness of a smart city project include infrastructure, data and services [22]. These specific areas should be the subject of evaluation in relation to their PEM assessment for a more proper and accurate evaluation of a smart city project. By doing so, for instance, project managers can evaluate objectives or resources of the subprojects in charge of ICT infrastructure and data-sharing platforms, respectively, so that they can address exactly the specific area in which the poor performance of the smart city project arose. This extended PEM evaluation could play a crucial role for smart city projects where different areas of the project are interconnected, and it becomes difficult to identify the source of many problems.

The active participation of expert groups dealing with smart city technologies in the project management process cannot avoid their involvement in the decision-making process and the assignment of responsibilities. This necessarily requires a development of a specific project governance that can adjust between the interests involved in the key components of a smart city and those of the PEM evaluation. The success of a smart city project management will depend on the development of such a governance model.

To secure visibility of the entire project's performance at a glance, Segway model matrix may be used. Items in the x and y axes can be expanded according to the project's characteristics. In the example below, on the X-axis, the Smart city's key components [19] are specified. On the Y-axis, the PEM evaluation elements are listed. This matrix aims to evaluate the performance of a smart city project based on Likert scale each evaluation value will be as follows: Very Unsatisfied = 1, Unsatisfied = 2, Neutral = 3, Satisfied = 4, and Very satisfied = 5.

In the existing PEM methodology, it is difficult to evaluate the complexity of projects involving various organizations and technologies due to the subjectivity of the evaluator. It is also difficult to divide all of the areas according to the field of expertise. Using this Segway model matrix, we can review the results of PEM evaluation according to each and every major component of a massive and complex project by identifying the excellence of subdivided components or areas that need further improvement, which will improve the performance and result of megaprojects (Fig. 13.3).

For a more detailed evaluation of a smart city project using the Segway model matrix, key performance indicators must be finely determined. The key indicators proposed by K-water [12] are designed to evaluate the expected performance of the Busan Eco Delta Smart City under these three main areas to be considered for a smart city technology, nature, and people. These areas further broken down into 35 specific indicators such as water quality management, flood forecasting, smart health care, sewage and heat energy, and e-government [12]. Our next research aims to analyse such key performance indicators and investigate what factors have to be considered for a performance evaluation depending on the type of smart city focusing on those currently being planned for construction worldwide.

SEGWAY MODEL MATRIX (IPMA PEM - SMART CITY SUCCESS FATOR)			SMART CITY SUCCESS FACTOR							
			A. Infrastructure			B. Data		C. Service		etc.
			A.1 Civil & Architecture	A.2 ICT	A.3 Geographic Information Syster	B.1 IoT	B.2 Data Sharing	C.1 Data analysis & Services	C.2 City Innovation Service	EXPANDABLE
IPMA PEM (International Project Management Association Project Excellence Model)	A. People & Purpose	A.1 Leadership & Value								
		A.2 Objectives & Strategy								
		A.3 Project team, partners & Suppliers								
	B. Processes & Resources	B.1. Project Management Processes & Resources								
		B.2. Management of Other Key Processes & Resources								
	C. Project Results	C.1 Customer Satisfaction								
		C.2 Project Team Satisfaction								
		C.3 Other Stakeholder Satisfaction								
		C.4 Project Results and Impact on the Environment								
	etc.	EXPANDABLE								

Fig. 13.3 Segway model matrix applying the key components of smart city

The 5-1 Life Zone in Sejong City and the Eco Delta Smart City project in Busan are still in the implementation stage, so it is difficult to prove the proposed model, but it is expected that it will be meaningful in terms of suggesting a new approach to project excellence evaluation in smart city. In order to verify the effectiveness of the Segway model, we would like to continuously update the study while the national pilot smart city projects are being carried out. At the same time, we will establish the methodology for designing the success factors for a smart city project and the examine effectiveness of the Segway model matrix based on these various success factors.

Acknowledgements This research was supported by the MOTIE (Ministry of Trade, Industry and Energy) in Korea, under the Fostering Global Talents for Innovative Growth Program (P0008747) supervised by the Korea Institute for Advancement of Technology (KIAT).

Ethics declarations

Conflict of Interest

The authors declare that they have no conflict of interest.

References

1. Bou-Llusar JC, Escrig-Tena AB, Roca-Puig V, Beltrán-Martín I (2009) An empirical assessment of the EFQM excellence model: evaluation as a TQM framework relative to the MBNQA Model. J Oper Manag 27(1):1–22
2. Camhi, J (2016) This one problem could cost smart cities hundreds of billions of dollars. In: The smart city report: forecasts, definitions, and strategies and takeaways from Barcelona, a leading smart city (Business Insider Intelligence, 2015)
3. Chaudhry S, Saxena S, Kumar D (2018) India's smart cities mission: smart for whom? Cities for whom?, update 2018. Housing and Land Rights Network, India, p 48
4. Cooke-Davies T (2002) The "real" success factors on projects. Int J Project Manage 20(3):185–190
5. Farahani MAJID, Sobhieh MH, Khameneh A (2009) Developing PEM using ICB standard to achieve integrity & relationship between them. In 5th International Project Management Conference, vol 4. Tehran, Iran
6. Flyvbjerg B (2017) Introduction: the iron law of megaproject management. In: Flyvbjerg B (ed). The Oxford Handbook of Megaproject Management, Oxford University Press
7. Hansen T (2017) News from smart cities: turning opportunity into reality, voices on infrastructure. GII, McKinsey & Company
8. International Project Management Association (2016) IPMA project excellence baseline, Version 1.0. International Project Management Association. Amsterdam, The Netherlands
9. IPMA (2020) Global Project Excellence Awards Large-&Mega-Sized projects. https://awards.ipma.world/about, Accessed 02 Feb 2020
10. Jang HY (2018) Smart city policy trends and future agendas in Korea. J Korea Acad Ind Coop Soc 19(6):507–518
11. Komninos N (2013) What makes cities intelligent? In: Deakin M (ed) Smart cities: governing. Modelling and Analysing the Transition, Talyor and Francis
12. K-water (2019) Busan eco delta smart city master plan, pp 28–34
13. Lessard D, Sakhrani V, Miller R (2014) House of project complexity-understanding complexity in large infrastructure projects. Eng Proj Organ J 4(4):170–192
14. Levitt R, Scott W (2017) Institutional challenges and solutions for global megaprojects. In: Flyvbjerg B (ed.) The Oxford Handbook of Megaproject Management, Oxford University Press
15. McKinsey (2017) Global Infrastructure Initiative (GII) Report
16. Ministry of Land, Infrastructure and Transport of South Korea (2013) The 2nd Ubiquitous City Master Plan. Available via DIALOG. https://smartcity.go.kr/. Accessed 02 Feb 2020
17. Park C (2016) Project ecosystem competency model. Proc Soc Behav Sci 226:116–123
18. Pinto JK, Prescott JE (1988) Variations in critical success factors over the stages in the project life cycle. J Manag 14(1):5–18
19. Presidential Committee on the Fourth Industrial Revolution (2018) Smart city promotion strategy. Smart City Prospects and Korea's Competitiveness
20. Rochet C (2018) Chapter 1: what do we mean by smart city and where does this idea come from?. In: Smart cities: reality or fiction. John Wiley & Sons
21. Scheiblich M, Maftei M, Just V, Studeny M (2017) Developing a project scorecard to measure the performance of project management in relation to EFQM excellence model. Amfiteatru Econ 19(11): 966–980
22. Joint Ministry (2018) Strategy and plan of smart city promotion, presidential committee on the forth industrial revolution
23. Westerveld E (2003) The Project Excellence Model®: linking success criteria and critical success factors. Int J Project Manage 21(6):411–418
24. Xia W, Lee G (2004) Grasping the complexity of IS development projects. Commun ACM 47(5):68–74
25. Yin RK (2017) Case study research and applications: design and methods. Sage Publications

Chapter 14
Public-Private-People Partnership (PPPP) for Infrastructure Development in Indonesia

Lukas Beladi Sihombing, Achmad Jaka Santos, and Andreas Wibowo

Abstract There is a continuous increase in the investment gap between government financing and infrastructure development. Therefore, to reduce this gap, the government used the services of the private sector to participate in infrastructure development under the public-private partnership (PPP). However, this scheme tends to often overlook the involvement of the surrounding communities around a project development area, with the ability to jeopardize the sustainability of the project. This research, therefore, proposes a new concept referred to as public-private-people partnership (PPPP). Questionnaires were used to collect data from a survey of 46 respondents. The result showed that the PPPP concept has the ability to benefit and support the livelihood of people living around the project area, thereby ensuring its sustainability.

Keywords Infrastructure needs · Public-private partnership (PPP) · Public-private-people partnership (PPPP)

L. B. Sihombing (✉) · A. J. Santos · A. Wibowo
Department of Civil Engineering, University of Pelita Harapan, Jl. MH. Thamrin Boulevard 1100, Klp. Dua, Karawaci, Tangerang, Banten 15811, Indonesia
e-mail: lukas_sihombing@yahoo.com

A. J. Santos
e-mail: jaka.santos@yahoo.co.id

A. Wibowo
e-mail: andreaswibowo1@yahoo.de

A. J. Santos
Postgraduate School, Djuanda University, Jl. TolJagorawi no.1, Ciawi, Bogor, Jawa Barat 16720, Indonesia

A. Wibowo
Agency for Research and Development, Ministry of Public Works and Housing, Jl. Panyawungan Cileunyi Wetan Kab, Bandung 40393, Indonesia

R. Cuevas et al. (eds.), *Research on Project, Programme and Portfolio Management*, Lecture Notes in Management and Industrial Engineering,
https://doi.org/10.1007/978-3-030-60139-3_14

14.1 Introduction

The global infrastructure investment gap is estimated between US$1 and $2 trillion annually. According to Ransdell [1], the annual investment gap in Asia is $200 to $400 billion, however, this analysis excludes China. In Indonesia, the government has estimated that the need for infrastructure investment till 2020 is approximately $169.7 billion, with its ability to only afford $15.1 billion, leaving a gap of $154.6 billion [2].

Bennon et al. [3], stated that the difference between estimated infrastructure spending requirements and the available investment capital from traditional sources are obtained from public pension funds [4], bank infrastructure [5], the private sector/regional initiatives [6] and hybrid of deep discount project bonds/land leases [7], most of which utilize the public-private partnership (PPP) scheme.

However, the use of this method has not achieved its optimal objective in filling the investment gap for infrastructure [8] due to lack of transparency in the PPP contract process [9]. Furthermore, society often failed to benefit from the scheme, because it lacked merits and positive pullover effects [10], with the concessionaire to accept its risk [12]. As a result of this, particular refinements of the PPP framework are needed that focuses on the indirect impact [13].

To obtain leverage effects, this study proposes the application of a public-private-people partnership (PPPP) scheme for infrastructure development in Indonesia. Therefore, this research addresses peoples' role in accepting the PPPP.

14.2 Literature Review

According to Guevara [8], PPP is the general terminology in the area of contracts and business, where the term "partnership" is defined as: "A voluntary association of two or more persons in carrying out a business for profit." Similarly, "public" is related to the process of involving an entire community, state or country. At the same time, "private" is the process of involving an individual, as opposed to the public or the government. Therefore, Guevara [8] defined PPP as a contractual agreement typically carried out by the association of a public and private party.

The London Underground PPP project, failed to achieve its aim because the contract had no provision for the direct control of taxpayers' cash flow to the company. According to Khoteeva and Khoteeva [14], there was an overestimated reliance of the government on the private sector's money. The failure was also due to the different project partner and socio-political factors. Soomro [15], stated that in many cases, there were public protests towards infrastructure privatization related to social welfare. This was also because the PPP project failed to reduce the desirable characteristics expressed in the purchasing process and fulfilment of the end-users expectations [16].

Kuronen et al. [17] reported that the "people" in PPP denotes a public-private-people partnership (PPPP), which offers an alternative approach for urban planning to establish interactions between the developer and the end-user. Therefore, they need to be both organizers and the guiding principle in every step of the program's development [18].

The people in PPPP are identified as local communities, NGOs, professional organizations, academic institutions, and media [19]. Furthermore, PPPP is a reinforcement of the grassroots economy, due to peoples' partnership with the government in the designing, planning and provision of architectural designs [20, 21]. The societal partnership has the ability to positively influence its citizens' engagement that results in public acceptance of the project and pressure on all organizations to complete it as soon as possible [11].

PPPP is currently being implemented in countries, such as in Finland (Helsinki), Nigeria, Hong Kong, China, India and Nepal. Table 14.1 shows further numerous possibilities used to examine PPPP.

14.3 Methodology

A survey was conducted to analyse the importance of people's role in PPPP. It was divided into a local Indonesian and global contexts. This was an online survey conducted from September 1 to 12, 2018 using a total of 46 respondents, as shown in Table 14.2.

The following questions were put forward during the survey: (1.) How important is the role of society (people) in financing infrastructure? (2.) Is the societal context more prioritized towards a direct effect with infrastructure development? (3.) Are the above-mentioned societal role in the following phase: land acquisition, planning, construction or operational? (4.) Are most of the lands in the area of indigenous communal property? (5.) Is the indigenous communal property used for infrastructure development? (6.) Is the indigenous communal property used for infrastructure development? (7.) In what form does the society play a role in infrastructure financing: private, state, village or special institution? (8.) Does society's involvement in financing improve its welfare around the infrastructure?

Next, the survey results attempted to devise the PPPP concept for the toll road infrastructure case in Indonesia.

14.4 Results

Based on the survey, the results showed that 39.1% respondent stated that "the role of society (people) in financing infrastructure is important". In addition, there were 71.7% respondent said "Yes" that the societal context here more prioritized towards society.

Table 14.1 The development of PPPP

No	Author(s)	Output	Role of people in PPPP	Country
1	Kuronen et al. [17]	To reduce carbon dioxide emissions	"a promising approach to decrease the carbon emissions	Helsinki
2	CTO [22]	Building ICT	To set up networks (land, labour etc.), as well as demand and utilize value added services	Nigeria
3	Thomas Ng et al. [23]	Infrastructure planning and policy formulation	To realize the changes associated with public aspirations and demands for infrastructure planning and policy formulation	Hong Kong
4	Zhang et al. [24]	A post-disaster reconstruction	"People" provide the "missing link" in traditional PPP to achieve effective and integrated partnership between multiple participants	China
5	Modi [25]	The policy of Narendra Modi	"Jan Andolan" (people's movement) with the total "Jan Bhagidari", including housing, health, education, and the adoption of orphan children or physical infrastructure	India
6	Acharya [26]	Building hydropower	The local people are urged to invest with the potential to receive shares of up to 24%	Nepal

Table 14.2 Respondent descriptions

Age (years)		Occupation		Education level	
< 17–25	6.50%	Civil servant apparatus, Army/Policeman, State-owned enterprise	34.80%	Undergraduate degree	27.90%
25–55	52.20%	Private company employee	32.60%	Master's degree	58.10%
> 55	41.30%	Entrepreneur	19.60%	Doctorate degree	14%
		Others	13.00%		

Table 14.3 Result of survey

No	Questions	Result
1	How important is the role of society (people) in financing infrastructure?	• 34.8% very important • 39.1% important • 23.9% enough
2	Is the societal context here more prioritized towards society, which has a direct effect with infrastructure development (the society who lives around the infrastructure development)?	• 71.7% said "'Yes" • 28.3% said "No"
3	Are the above-mentioned societal roles in the following phase: land acquisition, planning, construction or operational?	• 50% at land acquisition; • 19.6% at planning; • 10.9% at construction; and • 19.6% at operation
4	Are most of the lands in the area of the indigenous communal property?	• 26.1% said "Yes" • 17.4% said "No" • 56.5% said "Maybe"
5	Is the indigenous communal property used for infrastructure development?	• 8.7% said "Yes" • 21.7% said "No" • 69.6% said "Maybe"
6	In what form does society play a role in infrastructure financing?	• 45.7% in private-owned enterprises (cooperatives, limited companies); • 21.7% in State-Owned Enterprise/Region-Owned Enterprise/Village-Owned Enterprise; • 19.6% in Special Institution; and • Others 13%
7	Does society's involvement in financing improve its welfare around the infrastructure?	• 80.4% said "Yes" • 4.4% said "No" • 14.2% said "Maybe"

The survey also revealed that 50% respondent stated that "societal roles in the following phase are land acquisition". Otherwise, there were 69.6% respondent said "Maybe" that the indigenous communal property used for infrastructure development. Therefore, 45.7% respondent determined that society play a role in infrastructure financing in form of private-owned enterprises (cooperatives, limited companies). Finally, 80.4% respondent agreed to say "Yes" that society's involvement in financing improve its welfare around the infrastructure. More details, the survey results can be shown at Table 14.3.

14.5 Discussion

From the above survey, it is summarized that PPPP can be significantly applied to infrastructure building development in Indonesia. The peoples' roles are mentioned in the land acquisition phase, with their willingness to surrender their land for infrastructure development [20]. According to Zhang et al. [24], people need to create private-owned enterprises in order to work with public or private, professional organizations. Communities are, therefore, surrounded by the infrastructure building to improve their welfare, as stated by the Asia News Monitor [21].

In Indonesia, the use of toll toad infrastructure originated from PPP structure concept. Therefore, the State Asset Management Institute (SAMI) receives land acquisition funds that have already been allocated in the State Budget from the Ministry of Finance. Furthermore, it provides authority to the Toll Road Regulatory Agency (TRRA), based on the contracting agency through the Ministry of Public Works and Housing. The government, through the Ministry of Finance, provides Infrastructure Guarantee Fund (IGF) that offers a recourse agreement to the contracting agency of the toll road company (TRC).

For private organizations, after winning a toll road tender, the investor forms a special purpose vehicle (SPV) for the finance project concept and build-operate-transfer (BOT), on the TRC. According to the finance project concept, to acquire toll road construction funds, TRC obtains equity from a sponsor and debt from senior/junior/subordinated bank. After purchasing the land from the government, it proceeds to build toll road construction till the concession phase.

The surrounding community comprises of cooperatives, and limited companies, designed to manage the "people's" money from the results of the land usage for toll road infrastructure to its acquisition phase. The PPPP conceptual framework for building toll road infrastructure is shown in Fig. 14.1.

The conceptual framework of PPPP is a combination of the previously established PPP and people. Its advantage is to help the government in the execution of difficult and complex land acquisition transactions. Therefore, to ensure that the model is properly implemented, it needs a validation (e.g. in-depth interview with experts).

14.6 Conclusions

In conclusion, the PPPP conceptual framework is a viable solution used to fill the infrastructure investment gap. It is constructed from the existing PPP pattern and plays a significant role in contributing to social welfare, by building toll roads, which has a direct impact on infrastructure development.

The people's role in PPPP is recommended to take the form of private-owned enterprises such as cooperatives and limited companies. These organizations are established for communities to become legally protected, with incurred benefits from

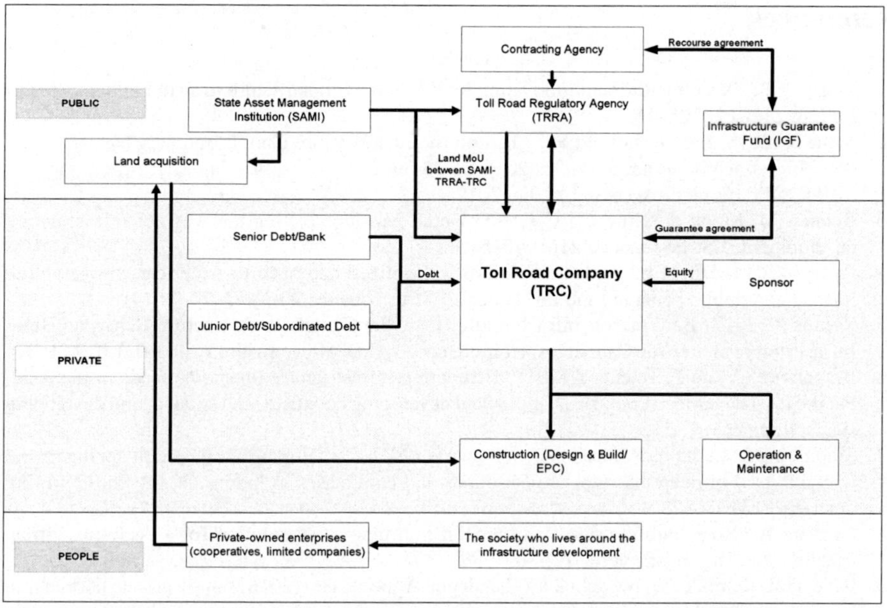

Fig. 14.1 Framework of public-private-people partnership in toll road infrastructure

the infrastructure building, to promote prosperity. The present research has demonstrated the potential of the proposed PPPP concept, however, it failed to address critical success factors and risk management, thereby, providing an avenue for future research.

14.7 Compliance with Ethical Standards

We declare that I ensured the objectivity and transparency in our research and that accepted principles of ethical and professional conduct have been followed. Prior informed consent was obtained from individual participants included in the study before the research. No sensitive personal data was accessed. Anonymity of individual participant data is maintained. And our research is not requiring an approval from the ethical body, as it is mentioned in the waiver 323/MTS-PPs-UPH/VII/2020, issued by *Universitas Pelita Harapan, Indonesia.*

References

1. Ransdell J (2019) Institutional innovation by the Asian infrastructure investment bank. Asia J Int Law 9(2019):125–152
2. Salna K (2018) Indonesia need $157 Billion for infrastructure plan. Bloomberg News. https://www.bloomberg.com/news/articles/2018-01-25/indonesia-seeks-to-plug-157-billion-gap-in-nation-building-plan. Accessed 21 Jan 2019
3. Bennon M, Monk A, Nowacki C (2015) Dutch pensions paving the way for infrastructure development. J Struc Finance 21(2):45–54
4. Ryan J (2014) Direct bilateral transactions: a localized approach to American infrastructure finance and public pension fund investment. J Struc Finance 20(3):57–70 (2014)
5. Robins S (2017) Banking on Infrastructure: How the Canada Infrastructure Bank can Build Infrastructure Better for Canadians. Commentary—C.D. Howe Institute, 483(0_1,0_2):1–31
6. Branchoux C, Lin F, Tateno Y (2018) Estimating infrastructure financing needs in the Asia-Pacific least developed countries, landlocked developing countries, and small island developing states. Econ 6(3):1–21
7. Sihombing L, Latief Y, Rarasati AD, Wibowo A (2018) Developing a toll road financing model with a hybrid of deep discount project bonds and land leases in Indonesia. Int J of Civil Eng Tech 9(6):1310–1323
8. Guevara A (2015) Public-private partnerships: an innovative solution for a declining infrastructure. The Urban Lawyer 47(2):309–337
9. Fadeyi OI, Kehinde OJ, Nwachukwu C, Adegbu Agboola OO (2018) Public private partnership for sustainable infrastructural development in lagos metropolis: prospects and challenges. Res Sci Today Marc 1:25–40
10. Mouraviev N, Kakabadse NK (2012) Conceptualising public-private partnerships. Soc Bus Rev 7(3):260–276
11. b. Mouraviev N, Kakabadse NK (2014) "Rules of engagement": how experiential learning facilitates the formation of a public-private partnership in Russia. J Mgmn Dev 33(6):551–563
12. Engel E, Fischer R, Galetovic A (2014) Risk and public-private partnerships. DICE Rep 12(3):3–7
13. Stadtler L (2016) Scrutinizing public-private partnerships for development: towards a broad evaluation conception. J Bus Ethics 135(1):71–86 (2016)
14. Khoteeva M, Khoteeva D (2017) Public-private partnerships: a solution for infrastructure development in the UK? Case study of the London underground public-private partnership project. Int Rev Mgmn Marketing 7(1):300–308
15. Soomro MA (2013) Failure mechanisms in transportation public private partnerships. Hong Kong University of Science and Technology, Hong Kong, Dissertations
16. Majamaa W, Junnila S, Doloi H, Niemistö E (2008) End-user oriented public-private partnerships in real estate industry. Int J Strat Property Mgmn 12(1):11–17
17. Kuronen M, Junnila S, Majamaa W, Niiranen I (2010) Public-private-people partnership as a way to reduce carbon dioxide emissions from residential development. Int J of Strat Property Mgmn 14(3):200–216
18. United Nations Economic Commission for Europe (UNECE) (2008) Guidebook on Promoting Good Governance in Public-Private Partnerships. ISBN 978-92-1-116979-9
19. Zhang J, Zou W, Kumaraswamy M (2015) Developing public private people partnership (4P) for post disaster infrastructure procurement. Int J Disaster Resilience Built Env 6(4):468–484
20. eHealth (2013) We Believe in Public Private Partnership with People, Noida (May 13)
21. Asia News Monitor (2016) Thailand: Progress of Government's implementation on grassroots economy reinforcement through people private public partnership. Bangkok, 19 Aug 2016
22. Commonwealth Telecommunication Organisation (CTO) (2011) Management of universal access funds for telecommunication regulatory board of Cameroon. https://www.slideshare.net/CTO-Research/management-of-usafs-module-six. Accessed 11 Sept 2018
23. Thomas Ng S, Wong JMW, Wong KKW (2012) A public private people partnerships (P4) process framework for infrastructure development in Hong Kong. Cities 31(2013):370–381

24. Zhang J, Zou W, Kumaraswamy M (2015) Developing public private people partnership (4P) for post disaster infrastructure procurement. Int J Disaster Resilience in the Built Env 6(4):468–484
25. Modi N (2018) Development through 4P model. https://www.narendramodi.in/people-public-private-partnership-3163. Accessed 11 Sept 2018
26. Acharya B (2018) Public private partnership in hydropower development: prospects for Nepal's development. https://www.slideshare.net/BikramAcharya/public-private-partnership-in-hydro-electricity-in-nepal-23324579?from_action=save. Accessed 11 Sept 2018

Chapter 15
Practice Competences in Project Management Decision Process: A Regional Study

Sandra Matuhina, Mladen Radujković, and Maja-Marija Nahod

Abstract The chapter focuses the decision-making process in project management from the perspective of the competence area "Practice", following IPMA ICB4., so as a behavioural pattern which is the foundation that project managers use while making decisions. There is no doubt about the importance of the practice competence elements like components of decision process. Therefore, an important research question is understanding the relation between practice competences, like defined by ICB4, and decision process in project management. The chapter brings preliminary results from a study conducted in three European countries: Croatia, Slovenia and Slovakia which share similar business environments, country characteristics, and cultural values (despite some differences). It highlights three main findings formed in the questions of traditional assumptions of decision-making. The main efforts of the study are oriented to the following: (1) the challenges of decision-making, (2) shifting the logic of the project management process to creative leadership, (3) the process of sustainable decision-making: advancing business and serving to the society. The chapter concludes with the main findings about the relevance of the competence elements from the aspect of competence area "Practice" according to IPMA ICB4 and specifically for the above-mentioned market, mostly on tied to certain differnces of infrastructure projects.

Keywords Decision-making · Project management · Practice competences · Infrastructure projects · Sustainability

S. Matuhina (✉)
Ekspertus Management Consulting,
Zagreb, Croatia
e-mail: sandra.matuhina@ekspertus.hr

M. Radujković
Alma Mater Europaea ECM, Maribor, Slovenia/Project, Program Portfolio Expert, Zagreb, Croatia
e-mail: mladen@projectexpert.hr

M.-M. Nahod
Ministry of Construction and Physical Planning, Zagreb, Croatia
e-mail: maja-marija.nahod@mgipu.hr

R. Cuevas et al. (eds.), *Research on Project, Programme and Portfolio Management*, Lecture Notes in Management and Industrial Engineering,
https://doi.org/10.1007/978-3-030-60139-3_15

15.1 Introduction

Decision-making is an important part of any project, program, or portfolio. However, despite it looks simple and logical, the decision process is always complex and sometimes claims for solutions far from the daily routine. The complexity of the decision process has many sources, since it brings together all key human aspects like competences, personality, information, interests, risks, culture, power play, experiences, … and many more. The heavy assignment usually forces project managers to find their own model to cope with decision-making, and professional associations like IPMA (International Project Management Association) recommend to relay on project management competences. There is no doubt about the importance of competent management at all levels up to the upper echelon level (top management) while dealing with projects. While observing from the management perspective, the decision-making process is a topic that permanently attracts increased interest in academic circles and also practitioners. Leadership is found to be significantly related to project management success, although there is no much of research of that kind, within the field of project management. Moreover, there are not too many studies about project management competences and decision-making processes, related to sustainability and benefits. In this context, there is requisite to understand what/how/why all echelons of project management do in a daily activity. To succeed, making decisions means to go far beyond the "iron-triangle" mindset, and claim for introducing competences and sustainability issues as important aspects of project management. This chapter explores project management from the behavioural economics perspective, particularly by focusing on project management—decision process—competences—project benefits at the large.

15.2 Objective

Without projects, we would not have business and society we do have today. From the historical perspective, we know that projects were a key driver of the development of various solutions for moving business, so as communities, to the next level. In parallel projects were an inspiration for many researchers while seeking a new management approaches and practices. Moreover, projects were also one of the main drivers and contributors in society development. It can be noticed that communities, which in the past accepted challenges and took initiatives to develop and implement successful projects moved their own community ahead in a much faster scenario. The wider objective of this study falls within the research question: what components of decision-making are the most significant to push a project forward close to success? Previous studies and literature review confirm the assumption that there is a space for research when it comes to decision processes within project management. However, as it is too complex a research scenario, we had to narrow it significantly. Therefore, the study was designed to focus decision-making in project management trough competence area “Practice” according to IPMA ICB4 [23], to check the foundation that project managers mainly use while making decisions. The preliminary research was conducted in three European countries: Croatia, Slovenia and Slovakia which share similar business environments, country characteristics and cultural values (despite some differences). Moreover, the study was limited to infrastructure projects. The main efforts of the study are oriented to the following: (1) the challenges of decision-making, (2) shifting the logic of the project management process to creative leadership, (3) the process of sustainable decision-making: advancing business and serving to the society. The chapter presents extracted preliminary results of the study.

15.2.1 The Power of Projects—Example of Infrastructures

A recent study on projectification confirmed the continuous growth of project activities within national economies [31, 34]. It expresses the need for changes, adaption and development, because projects are tools for achieving it. Last three decades academic community focused on different aspects of project management priorities and developments, such as outcomes, complexity [21], performance [4, 39], inceptions [17, 18, 26], etc. But, even under the best of circumstances, it is difficult to find valid and reliable data on the performance of particular types of projects, for example, infrastructure investments [1]. Recently, the profession of project management became much more interested in the benefits that infrastructure projects deliver. Project success is a complex issue, due to many perspectives and interest which are on the agenda while making an evaluation. However, researchers agreed that at the bottom level, there are two perspectives: the short term during the project close-out phase and long term during the project delivery usage and operations, where benefits from projects are delivered. [15]. Infrastructure projects usually have a long life cycle, high investments, consumption of many different resources, and pretty complex stakeholder matrix. In many cases, a particular infrastructure megaproject engages almost all capital of the community, while expected benefits are coming over a longer period. Research from PwC [30] shows that worldwide, infrastructure spending will grow to more than $9 trillion per year by 2025. It is expected to be spent on a global scale close to $78 trillion between 2014 and 2025. There are perceptions that military budgets are the top spending item while aggregating them at the global level, but for sure total global investment in infrastructures could compete with. The hunger for infrastructures is a result of the need for improving and advancing human life, and it is valid for developed, developing and not developed countries. The only difference is that emerging markets are speeding up their infrastructure development plans, while trying to fulfill requirements for economic growth and for improving the life quality for their citizens. However, infrastructure projects have other important impacts on society, like findings that for every one billion dollars of investment in infrastructure, as much as 20,000 new jobs can be created [25, 41]. So, infrastructure case is just one example and evidence of a power of projects while creating a community's better future. And each of those projects is managed by people exposed to high pressure for achieving planned objectives and for delivering success. Decision-making is their daily lifestyle, sustainability is "the light in the tunnel", competences are "the best friend" and people expectations are driver for their headache [37].

15.2.2 Literature Review on Project Management, Decision-Making and Leadership

While observing projects and people in charge, there are many roles and responsibilities. Probably one of the most challenging is the role of project manager. In its simplified version those people manage the project delivery process, either for predefined new products or predefined service. Their daily job integrates creative thinking, communication and decision-making in original scenario for each project they manage. In their daily job each of them brings their competences, positive practices from the past, personality and behaviour, interests and believes,..., so as other variables which form particular management pattern, and which in turn influence the success of their work. There are many studies about the management patter, particularly related to manager versus leader, or manager + leader. Many researchers addressed that leadership and its elements may have great importance on the outcomes of the project [29, 38, 40]. Some authors believe that such person should be visionary, integrator, agent of change, but the critical component is leadership which influence practice and contribute in development of a learning organization [24]. Other authors also believe that leadership is one of the most important competences for project manager, and that he/she might expect a lot of challenges from that side, while dealing with major projects [11, 27]. According to IPMA, competences in project management means application of knowledge, skills and abilities for achieving the programmed results [23]. Leadership in project management is focused on having a vision about the strategy, scope, objectives, team and stakeholders. In parallel, it claims for understanding the "big picture", which define the context where project will be realized. According to IPMA, the third element of the competence triangle is practice or technical competences, which represent the knowledge, methodologies, tools, techniques, within the project management profession. Successful project manager merges all above and make fine balance while directing the project towards its objectives and stakeholder expectations. Therefore, successful project execution [12, 28, 43] depends on the qualification of project manager, and change one person to another person at such position, will result with different outcomes. Successful project manager, which is in parallel also leader, has vision, or even feeling, how to bring all project and management elements together, and share it with his/her team like composition for success. There are many studies about the relation between performance and styles of leadership [5, 22]. However, much more demanding issue is how to incorporate and facilitate leadership aspect in project management. Some project managers are very eager to use the power of position, and to put his/her ego in the front. Some scholars recommended that the negative effects could be downsized by the training of leaders with different skills [9, 20, 32]. Project organizations often request leadership for their actions and decisions [2, 16]. Within specific business entity or organization, overall business and activities are divided and structured according to internal organizational scheme. Therefore, we can identify a lot of decision/leadership points or centers, which are connected in a hierarchical order. They form a system, similar to nerve system in a human body, which manage with parts and entire system at large. Even simple insight in such system could mark upper leadership levels like dominant decision centers, which highly influence overall system and its function. And project managers must fit in such reality.

15.2.3 Challenges for Decision-Makers and Upper Echelon of Infrastructure Projects Management

World's prominent infrastructure statistics gives evidence that there is a raising need for investing in infrastructure. Infrastructures serve to large number of people in particular community, so they have strong stakeholder support from different groups. There are different forecasts, even those which state that spending on infrastructures is the biggest item in the world budget, even higher than military and defense budgets (2 trillion USD per year). It is mainly related to boom of global population and need for certain life standard where infrastructures contribute with large share. Global strategist, Parag Khanna stated [19] wrote that the current global infrastructure capacity is at the level which could satisfy the needs only for three billion of people. And while moving from seven billion total world population to the nine billion, we should spend 1 trillion of USD per each 1 billion of people. One of the most important issues of infrastructure is to ensure enough finances and in parallel to ensure that such huge money is spent in the best possible way to produce the benefits for people [14]. Infrastructure projects are famous about cost and time overruns; therefore, they have great potential for the management and for the questioning for each unit of currency spent. While observing such scenario from the perspective of upper management, several key challenges could be immediately noted. The simple question could be formulated to the learning if there is right time for right project and right people to deliver it within the specific context. Speaking about infrastructures, each time looks like right time, and moreover seems we are always behind the needs. Context introduces external components, and sometimes limited space for maneuver, but nevertheless, this is responsibility of the upper management. And while speaking about infrastructures, right project is mainly question about the priorities, which means that it is selection one among many, during multi-criteria analyses, under the pressure of public and different interest groups, political programs and hidden agendas. There are many questions which need answers, such as lessons from the past, post-evaluations and findings, regulatory environment and its changes, monitoring of social an environmental impact, ..., so as numerous factors coming from people, business and planet perspectives. Everything above is under the domain of the upper echelon of management, and once they put it together, there is still one question left: do they have people to design and implement it, like defined? How to achieve the best possible success scenario? After considering it, they shift a part of decision process and responsibilities to the lower levels of management, represented in the role of project manager, claiming for the "success". While the upper management declare the success criteria, project manager has to investigate, declare and practice success factors across the project organization and project processes [44]. So, both, upper and execution level of management are involved in decision process, they are inter-dependable decision makers in permanent interaction with many interested groups (Fig. 15.1).

We could say that a successful project is one that brings value regarding its objectives and that in parallel it answers societal needs in its long-term outcomes [33, 35]. The Department for Transport in the UK published certain steps for the appraisal of benefits in large projects [10]. Four kinds of effect were identified as important: effects of productivity, effects of agglomeration, effects of competition, and labor-market effects. Obviously, project success is a hot topic for both researchers and practitioners, and a lot of progress has been made in raising that success to the next level; we have many macro-, mezzo- or micro-aspect studies dealing with this. However, there is no magic formula for achieving it, and there will be no such formula in the future because of the high level of complexity in all aspects of such projects and the big differences between them. But we can use previous findings, tips and results to create and tailor a successful scenario that accommodates the criteria

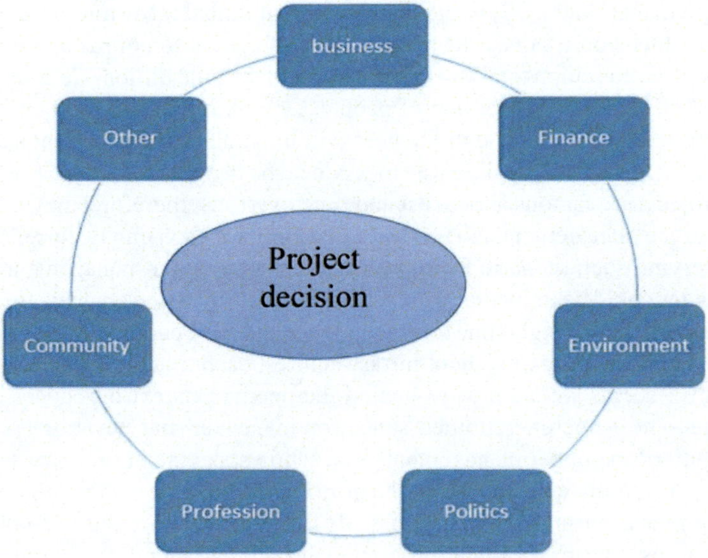

Fig. 15.1 Project decision and participating groups

and factors of importance in a specific case. While dealing with such assignment we should focus the following like foundation for success:

- linking project to proven community needs, and correspondent strategy
- alignment and active participation of all relevant stakeholders, and forming firm project coalition for project success, based on solid business case
- engaging of appropriate and sufficient competences and creativity to drive project concept to reality, and deliver end product
- assuring sustainability, that there is balance among business, people and planet factors trough each level of decision-making process.

And further contribution could be done through developing competences of PM profession, and new adaptive tools and methods for improvements, such as identifying risk management practices, front end planning impact evaluation, so as other. More efficient use of project management techniques and methods might improve project performance and reduce the risks so often related to infrastructure projects like cost overrun.

15.2.4 Sustainability Approach in Project Management Decision Process

In September 2015, the UN General Assembly adopted the 2030 Agenda for Sustainable Development that includes 17 Sustainable Development Goals (SDGs) which are crucial for the global human community [42, 45]. It is not only about individual or organization, it is about our world like we know. Considering that projects are represented in about 30% of human activities, especially within the economic, growth and change for better [34], it is important that each project incorporate sustainability, so as that project managers practice it. It means that besides the standard criteria of project success which represent different stakeholders and their interests, each project must have sustainability like criterion which represents global community welfare. Consensus among project participants is important concept of sustainability, and it is considered to be a complex concept [8]. Elkington [13] proposed a basic model for sustainability and included into the concept 3 elements: people, profit and planet. Silvius et al. [36] extended the sustainability view on project and project management aspects proposing the very specific aspects:

1. harmonizing economic, environmental and social interests
2. including long term and the short term
3. global and local
4. consuming income, not capital
5. accountability and transparency
6. ethics and personal values.

The OMEGA Centre team declare that effective development of project depends deeply on the stability of institutional frameworks [3]. The Green Projects "5P concept" considers that the sustainability of a project comprises governance and project management aspects but also the final product and its sustainability [6, 7]. Such sustainability-based project delivery method incorporates product and process elements to manage the balance between social responsibility, resources and delivering "green" project outcomes [7]. Whichever is decision-making model it must rely on sustainability, since it is foundation for long-term perspective success and ground where all stakeholders make compromise about the project scope and objectives. It means that project owner must include it in a project concept and project business case, while managers must take care that sustainability is part of all project relevant documents and success criteria. On the operational level, sustainability should be a glue which hold together all competence elements that project manager practice.

15.3 Methodology and Research

As declared above competences and creativity are one among basic parts needed for directing project towards success. Infrastructure projects engage many experts and disciplines, and project management profession is linking and directing factor. It is not just glue, but rather a living organism. In our research, we decided to collect sample and provide insight about project management competences related to the area practice, following IPMA ICB4 standard [23]. The final sample was composed by replies of 62 IPMA certified respondents who were mainly managing different infrastructure projects in Slovakia, Slovenia and Croatia, like three smaller EU countries, sharing similar cultures and business environment. The respondents were asked about the competences of decision-making process as part of project management process. There were 48 male and 14 female respondents. The structure of the respondents' education level points to the dominant higher education, while in the sample are also present individuals with a PhD or MBA, and at the least represented category, undergraduate level of education. Most respondents are currently engaged on a project. To a lesser extent, respondents pleaded that they were working on a program and the least represented was a certain portfolio of jobs they are currently working on. Respondents were asked about the estimated budget for the project they manage, and the average in the sample was $ 15,884,296.29 million. The sample has high standard deviation, which indicates that the budget data distribution has an extremely volatile average value that is not representative. The two most active sectors are construction (29.03%) and the public sector, education and health (27.42%). As the least represented industrial sector, production stands out (6.45%). Furthermore, the total sum of budgets per industry was analysed, and the average value of the budget for each industry, and the results of the analysis are shown in Fig. 15.2.

Although the largest sum of budgets is present in the construction industry, the average value of the budget is highest in financial and insurance services. The average value of the budget exceeds the sum of values in the production sector. The exact values of the budget sum and average are shown in Table 15.1.

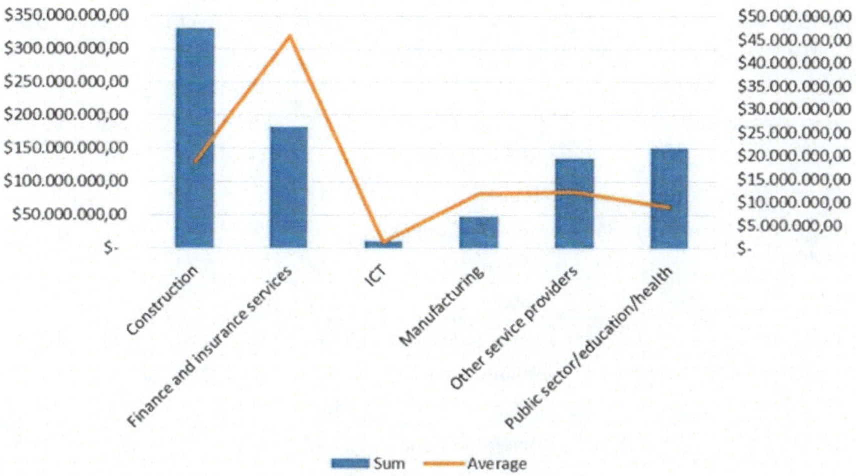

Fig. 15.2 Combined graph of industry sectors, sums and averages of budget values in $ *Source* authors

Table 15.1 Budget sums and averages by industry sectors

Industry sector	Sum	Average
Construction	$331.526.400,00	$18.418.133,33
Finance and insurance services	$182.400.000,00	$45.600.000,00
ICT	$ 11.004.000,00	$ 1.375.500,00
Manufacturing	$ 47.233.600,00	$11.808.400,00
Other service providers	$134.520.000,00	$12.229.090,91
Public sector/education/health	$151.068.000,00	$ 8.886.352,94

For examination of the differences between gender categories in decision-making process, the t-test of mean difference was conducted. Results are presented in Table 15.2.

Results indicate that there are statistically significant differences between genders in assessment of practical competences, were the female population perceives in average these competences, more important than male population ($t = -2.78$; $p <$

Table 15.2 T-test of mean differences between genders in decision-making process

Variable	Category	N	Me	Sd	t	p
Practical Competences	Male	48	76.27	8.34	−2.78	<0.05
	Female	14	82.00	6.26		
Decision-making process	Male	45	63.18	9.53	2.36	<0.05
	Female	13	55.92	9.84		

Table 15.3 F-test for multi-group comparison (years of experience, current occupation, country and IPMA level) for variable 'practical competences'

Variable		SS	df	MS	F	p
Years of experience	Between Groups	31.65	28	1.13	0.78	>0.05
	Within Groups	47.52	33	1.44		
	Total	79.18	61			
Currently working on (portfolio, program, project)	Between Groups	11.05	28	0.39	1.12	>0.05
	Within Groups	11.66	33	0.35		
	Total	22.71	61			
Country	Between Groups	6.07	28	0.22	0.74	>0.05
	Within Groups	9.61	33	0.29		
	Total	15.68	61			
IPMA Level	Between Groups	27.33	28	0.98	1.44	>0.05
	Within Groups	22.41	33	0.68		
	Total	49.74	61			

0.05). When asked about perception and importance of decision-making process, the male population perceives this process statistically more significant than female population ($t = 2.36$; $p < 0.05$).

Further multi-group test was conducted to try to detect statistically significant differences between observed groups. For this purposes *F-test* for multi-group comparison was conducted. There were no statistically significant differences between observed groups. Results of the analysis are presented in Table 15.3.

Multi-group analysis was conducted to analyse differences in variable "decision-making process", and the results were similar as with previous variable. Results are presented in Table 15.4.

15.4 Results and Discussion

The following chapter contains main information about the essence of the results. Figure 15.3 shows the most important competences of practice elements as marked by respondents, and as referring to IPMA ICB4 standard. Research showed that 3 out of 14 practice competences showed higher appearance than others, and those were: plan & control, finance, so as requirements & objectives & benefits. While comparing to our previous studies done during the last ten years, we found that planning and control, so as finance is always focused by PM profession in the region. It could mean that project managers are continuously struggling with time and cost perspective of projects, due to overoptimistic initial goals or shortages of money/time. Element requirements and objectives point to potential difficulties regarding what project stakeholders want to achieve and what project accomplish. By linking all

Table 15.4 F-test for multi-group comparison (years of experience, current occupation, country and IPMA level) for variable "decision-making process"

Variable		SS	df	MS	F	p
Years of experience	Between Groups	41.897	33	1.270	.967	>0.05
	Within Groups	31.500	24	1.313		
	Total	73.397	57			
Currently working on (portfolio, program, project)	Between Groups	11.013	33	0.334	0.737	>0.05
	Within Groups	10.867	24	0.453		
	Total	21.879	57			
Country	Between Groups	9.551	33	0.289	1.164	>0.05
	Within Groups	5.967	24	0.249		
	Total	15.517	57			
IPMA Level	Between Groups	26.837	33	0.813	1.322	>0.05
	Within Groups	14.767	24	0.615		
	Total	41.603	57			

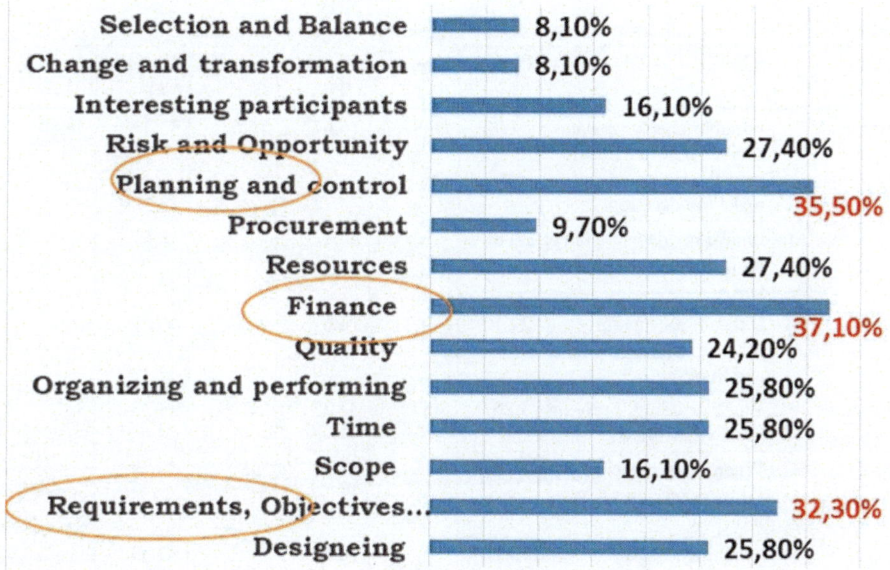

Fig. 15.3 Structured columns of the most important competences perceived by participants

three elements, seems could be no balance among project stakeholder expectations—time—money in the project, and such scenario is burden for the PM profession. On the other hand, there is a set of competence elements which include risk and opportunity, participants/stakeholders, resources, quality, time, scope which are standard elements for most of project types. This confirms that project management profession in the region is aligned with international practice and standards in project management. Element design is much known to all those dealing with infrastructure projects. No matter how infrastructure project is designed, and who and how it did, project manager can always expect some issues from that side. Taking all together elements which are above twenty percentages make scenario which is familiar for project manager's experienced in (particularly infrastructure) projects management: risk & opportunities, plan & control, resources, finance, quality, organization & information, time, requirements & objectives & benefits, design. Seems that those elements or topics dominate in daily project manager work, while making decisions.

In Table 15.5 there are main results of the decision-making process, distributed in the three categories: rational, intuitive and spontaneous. Out of observed answers survey participants demonstrated pretty balanced decision-making model. The overall results in the table below, do not show harmony among the countries, but total average taken is 28% for rational model, 23% for intuitive and 28% for spontaneous. The participants from Croatia reported low presence of intuitive and spontaneous behaviour, and we believe that it does not match the actual situation. According to our opinion, the results from Slovakia and Slovenia represent the regional pattern in

Table 15.5 Decision-making process is divided into three separate categories

Category	Item	Croatia (%)	Average (%)	Slovakia (%)	Average (%)	Slovenia (%)	Average (%)
Rational	I double-check my data sources before I'm sure I have the right facts to make my decision.	34.00	28.97	16.67	16.67	33.33	38.89
	I make my decisions in a logical and systematic way.	30.80		N/A		16.67	
	My decision-making process requires caution and thinking.	22.10		16.67		66.67	
Intuitive	When deciding, I rely on my own instincts.	3.80	7.57	33.33	27.78	33.33	33.33
	When I make decisions, I always rely on my own intuition.	5.70		33.33		33.33	
	Generally, I make decisions that feel right.	13.20		16.67		33.33	
Spontaneous	I often make a quick decisions.	1.90	2.53	16.67	27.78	33.33	55.55
	I often make decisions triggered by the moment.	3.80		16.67		33.33	
	When making decisions, I do what seems natural at a given moment.	1.90		50.00		100.00	

much better way, since they show more intuitive and spontaneous decision-making cases. However, before the survey, we expected even higher score of the spontaneous decision-making. Such expectation was related to the culture in region, so as heritage from the past. In addition, in the most organization (particularly public owned) there is still no so rigid business system with detailed and prescribed procedures, so an individual faces situation where he/she must act with great degree of own perception and initiatives. On the other hand, culture in the region support inspiring and creative acting, where an individual has enough space for creativity. It is expected that future will bring increase in rational based decision-making situations, due to adopting project management methodologies, practices and procedures. Many organizations are keen to introduce framework for success facilitation, where proven methodologies and best practices provide guideline for decision-making in a complex project situation. However, projects are non-routine undertakings, with limited similarity and no chance for copy-paste. Therefore, an organization must not build rigid business environment, which prevents creativity in decision process. In that way, certain share of intuitive and spontaneous decisions will be normal part of the decision package.

The main limitation of the study is related to the sample, which is combined and rather small, so results are preliminary and informative. For the future research, we recommend participation of more respondents and maybe selective approach by sectors, since projects have different challenges and features in different sectors, also budget characteristics even if all falls within some category of infrastructures.

15.5 Conclusion

Projects have many "?" across all features and phases, always linked with some kind of participation and expectations. While some of them have more commercial foundation and expectations, the other have more community-related expectations (i.e. for particular infrastructures). Managers are exposed to enormous amount of data, challenges, expectations and risks in their daily work. Decision-making is their daily life style, sustainability is "the light in the tunnel", competences are "the best friend" and people expectations are driver for their headache. In this study, preliminary results showed that project managers marked plan & control, finance, so as requirements & objectives & benefits as the most significant practice competence for the region of three selected countries, Slovakia, Croatia and Slovenia. Balanced decision-making is stated as preferred leadership style, which brings together all observed elements to success. Although leadership topic is covered in management science in general, there is lack of regional research on this topic specifically on projects. Sustainability, competences and decision process are getting more importance by the profession of project management. In this research Croatian, Slovenian and Slovakian individuals and companies showed certain interest for the study which brings together all those items. Therefore, it is recommended to continue research on much wider sample and trough different scenarios.

15.6 Compliance with Ethical Standards

Informed consent was obtained from the participants, meaning that they were informed about the research objective and research methodology. They agree to contribute to the research. Collected data was anonymised and special procedures to assure anonymity were applied. The study reported in this paper was authorized by the Project Management Association of Slovakia under permission reference 001-2017, the Project Management Association of Slovenia under permission reference DOP_65/2017 and Project Management Association of Croatia under permission reference 02-88-2017.

References

1. Ansar A, Flyvbjerg B, Budzier A, Lunn D (2016) Does infrastructure investment lead to economic growth or economic fragility? Evidence from China, Oxford Review of Economic Policy, vol 32
2. Beatham S, Anumba CJ, Thorpe T, Hedges I (2004) KPIs—a critical appraisal of their use in construction. International Journal of project management, vol 11
3. Beynon M, Curry B, Morgan P (2000) The Dempster-Shafer theory of evidence: an alternative approach to multicriteria decision modelling, Omega, vol 28
4. Brookes N (2015) Mankind and Mega-projects, Frontiers of Engineering Management
5. Bycio P, Hackett RD, Allen JS (1995) Further assessments of Bass's conceptualization of transactional and transformational leadership. J Appl Psychol 80
6. Burcar Dunović I, Łukasiewicz A, Brookes N.J (2014) In pursuit of a framework to explore stakeholder influence in megaproject, in People, Buildings and Environment 2014, Presented at international scientific conference, Kroměříž, Czech Republic
7. Carboni J, González M, Hodgkinson J (2013) PRiSM™—Projects integrating Sustainable Methods: The GPM Guide to Sustainability in Project Management
8. Collinge B (2012) Re-thinking Stakeholder Management in Construction: Theory & Research, Project Perspectives, vol XXXIV
9. Darcy, T., Kleiner, B.H.: Leadership for change in a turbulent environment, Leadership Organisation Development Journal, vol 12 (1991)
10. Department for Transport: Transport, Wider Economic Benefits and Impacts on GDP, Department for Transport, London, U.K. (2005)
11. Dijksterhuis A, Bos MW, Nordgren van Baaren RB (2006) On making the right choice: the deliberation-without-attention effect. Science 311
12. Dulaimi MF (2005) The influence of academic education and formal training on the project manager's behaviour. J Constr Res 6
13. Elkington J (1998) Cannibals with forks: the triple bottom Line of 21st Century Business, New Society Publishers
14. Estache A (2010) A Survey of Impact Evaluations of Infrastructure Projects, Programs and Policies, Working Papers ECARES, Université Libre de Bruxelles
15. Fahri J, Biesenthal C, Pollack J, Sankaran S (2015) Understanding megaproject success beyond the project close-out stage. Construction Economics and Building 15
16. Fiedler FE (1996) Research on leadership selection and training: one view of the future. Administr Sci Quart 41
17. Flyvbjerg B (1997) the aalborg study: case selection and dana selection. Aalborg University, Department of Development and Planning, Aalborg

18. Flyvbjerg B (2007) Cost overruns and demand shortfalls in urban rail and other infrastructure. Transportation Planning and Technology 30
19. Futurist Speaker Blog: Parag Khanna (2016), http://www.futuristspeaker.com/job-opportunities/megaprojects-set-to-explode-to-24-of-global-gdp-within-a-decade/. Accessed 15th August 2019
20. Hennessey JT (1998) Reinventing government: does leadership make the difference? Public Adminstr Rev 58
21. Hertogh M, Westerveld E (2010) Playing with complexity: management and organisation of large infrastructure projects, Erasmus Universiteit Rotterdam
22. Howell JM, Avolio BJ (1993) Transformational leadership, transactional leadership, locus of control and support for innovation: key predictors of consolidated-business-unit performance. J Appl Psychol 78
23. IPMA: Individual Competence Baseline ICB4, IPMA (2015)
24. Jalava U, Virtanen P (2000) Innovatiiviseen projektinjohtamiseen. Helsinki, Tammi
25. Miller R (2013) Infrastructure: investment in jobs. Industrial Heating 81
26. Miller GA (1956) The magical number seven, plus or minus two. Some limits on our capacity for processing information. Psychol Rev 24
27. Morris PWG, Hough GH (1987) The anatomy of major projects. Wiley, New York
28. Munns AK, Bjeirmi BF (1996) The role of project management in achieving project success. Int J Project Manag 14
29. Odusami KT (2002) Perceptions of construction professionals concerning important skills of effective project leaders. J Manag Eng 18
30. PwC: Capital project and infrastructure spending Outlook to 2025, Research by Oxford Economics (2014), https://www.pwc.com/gx/en/capital-projects-infrastructure/publications/cpi-outlook/assets/cpi-outlook-to-2025.pdf. Accessed 14th May 2019
31. Radujkovic M, Misic S (2019) Projectification of economy in a smaller country: a case from Croatia. Int J Bus Manag Technol 3
32. Saari LM, Johnson TR, McLaughlin SD, Zimmerle DM (1988) A survey of management training and education practices in US companies. Personnel Psychol 41
33. Samset K (2013) Strategic and tactical performance of mega-projects—between successful and inefficient success. In: Priemus H and van Wee B (2013) International Handbook on Mega-Projects, Edward Elgar Publishing Limited, UK (2013)
34. Schoper YG, Wald A, Ingason HT, Fridgeirsson TV (2018) Projectification in Western economies: a comparative study of Germany, Norway and Iceland. Int J Project Management 6
35. Shirazi B, Langford D, Rowlinson S (1996) Organizational structures in the construction industry. Constr Manag Econ 14
36. Silvius G, Schipper R, Planko J, Köhler A, Van Den Brink J (2012) Sustainability in Project management. Ashgate Publishing Limited, UK
37. Suikki R, Tromstedt R, Haapasalo H (2006) Project management competence development framework in turbulent business environment, Technovation 26
38. Toor SR, Ogunlana SO (2006) Successful project leadership: understanding the personality traits and organizational factors. In: Proceedings of CIB-W107, International symposium, construction in developing economies: new issues and challenges, Chile, Santiago
39. Toor SR, Ogunlana SO (2008) Critical COMs of success in large-scale construction projects: evidence from Thailand construction industry. Int J Project Manag 26
40. Toor SR, Ofori G (2006) In quest of leadership in construction industry: new arenas, new challenges! In: Proceedings of joint international conference on construction culture, innovation, and management (CCIM), Dubai, UAE
41. United Nations (2016) World Investment report. United Nations, Geneva
42. Vickerman RW (2007) Cost–benefit analysis and large-scale infrastructure projects: state of the art and challenges, Environment and Planning B 34
43. Westerveld E (2003) The project excellence model: linking success criteria and critical success factors. Int J Project Manag 21

44. World Economic Forum (2014) Strategic Infrastructure Steps to Operate and Maintain Infrastructure Efficiently and Effectively. Prepared in collaboration with The Boston Consulting Group, World Economic Forum, Switzerland
45. UN. The 2030 Agenda for Sustainable Development (2015)

Chapter 16
From Social Responsibility to Sustainability: Are Excellence Awarded Companies in Iceland Socially Responsible?

Asa Bjork Jonsdottir, Haukur Ingi Jonasson, Helgi Thor Ingason, and Agnes Holm Gunnarsdottir

Abstract Excellence is a managerial virtue that is sought after within the world of commerce, and in many countries excellence awards in management are celebrated. Iceland is a country that is internationally recognized as being in the forefront in terms of social responsibility, equality and sustainable development. What are the best Icelandic companies doing with regard to sustainability in their portfolio, project and programme management? Icelanders can claim/boast two management excellence awards, *Outstanding Companies Award* and *Exemplary Company Award*. Although the awards' names may sound comprehensive, each award only looks at a limited aspect of the organization that is being scrutinized. This paper examines four international business excellence models with a view to illustrating how a company's excellence can be measured. All these excellence models emphasize *social responsibility* and sustainability and claim that an organization cannot truly be excellent without being up to date with regard to the urgent developmental issues we face. Based on this, it is surprising that when recipients of the above-mentioned awards are asked, it emerges that the excellence awards focus on stakeholders and do not challenge organizations to show excellence with regard to social responsibility and sustainability, or at best only to a very limited degree.

Keywords Sustainability · Project management · Excellence awards

A. B. Jonsdottir · H. I. Jonasson (✉) · H. T. Ingason · A. H. Gunnarsdottir
Reykjavik University, Menntavegi 1, 101 Reykjavik, Iceland
e-mail: haukuringi@ru.is

A. B. Jonsdottir
e-mail: asaj07@ru.is

H. T. Ingason
e-mail: helgithor@ru.is

R. Cuevas et al. (eds.), *Research on Project, Programme and Portfolio Management*, Lecture Notes in Management and Industrial Engineering,
https://doi.org/10.1007/978-3-030-60139-3_16

16.1 Introduction

A few years ago, one of the authors of this paper stumbled upon a sign that had been sprayed on a sidewalk in Melbourne, Australia stating: "*Iceland is the freest country on the planet!*" (Australia is on the exact opposite side of the globe from Iceland)—an excellent reputation indeed, at least for the liberally attuned. Furthermore, Iceland, along with the other Nordic countries, has been seen as a country which lays emphasis on socially responsible business, gender equality and sustainable development. Moreover, it seems to symbolize the aspiration towards key human values. If this is actually the case, it might be interesting to ask if Icelanders stand for something universally meaningful in how they go about things. So what can we learn from the Icelandic companies that have been identified as excelling at management in our search for insights that might have universal implications for the project management profession?

According to the Oxford Dictionary, excellence means: "the quality of being outstanding or extremely good". But what makes a business excellent? Quite a few models claim to measure business excellence. In this paper, I will be taking a closer look at four internationally recognized business excellence models. In the book "Afburðastjórnun", the authors investigate three excellence models and discover that they have a lot in common, e.g. they all measure success as having met or exceeded the expectations of four key stakeholder groups; the clients, the employees, the owners and society, for at least three consecutive years [1]. In Iceland there are a few awards that reward companies for performance excellence in a specific area. What do those awards mean, and is a company with an award more excellent than a company without one? One of the Icelandic awards is given by the company Credit Info and is the Outstanding Companies Award (Icel. Framúrskarandi fyrirtæki). According to the company's website, outstanding companies build their operations on solid ground and grow the interests of investors and shareholders. Its main purpose is to reward companies for doing good and contributing to a better business environment [4]. To be an Outstanding Company, nine financial figures or financially related measurements must meet a standard set by Credit Info [3] If we were to compare it to the business excellence models, it would fit nicely into the "expectations of owners" part.

Another award is the Exemplary Company Award given by the labour union VR. It measures nine key factors in the employees' work environment, management, workplace morale, salary, work conditions, flexibility, independence at work, company image, happiness and pride and lastly, equality. A survey is sent out to the employees, and the award is given to those fifteen companies which get the highest score in each of the three predetermined size groups [30]. Again, if we were to compare this award to the business excellence models, it would cover the "expectations of employees" part.

But is getting either of these awards enough for an organization to be truly excellent? An organization which gets both awards can be said to have met and/or exceeded the expectations of two of the four key stakeholder groups. What about the other two? Corporate Social Responsibility (CSR) awareness is on the rise, and as our society becomes even more transparent, the image of an organization can quickly be destroyed with a few pictures or quotes on social media. There are signs that this trend will only increase as a CSR survey done in 2017 by Cone Inc. finds that 79% of Americans expect companies to keep improving their CSR efforts. 87% also said that they would buy a company's product if it advocated an issue they cared about. 92% say they have a more positive image of a company when it supports a social or environmental issue, 87% are more likely to trust the company and 88% are more loyal to it. On the other hand, 88% would stop buying products from a company if they learned it was irresponsible or deceitful, and 50% reported that in the past 12 months, they had done just that [5].

In this paper, we investigate how the companies that have received both of these awards meet the fourth criterion of the excellence models in an effort to try and answer the question: Are "excellent" companies in Iceland socially responsible?

16.2 Literature Review

This section briefly describes four internationally recognized business excellence models; the Baldrige Excellence Framework, the EFQM Business Excellence Model, the Canadian Framework for Business Excellence and the Shingo Model. It will highlight what they have in common and what they can tell us about measuring the excellence of an organization. It will then briefly go into the topic of Corporate Social Responsibility (CSR) and how it can be measured.

16.2.1 The Baldridge Excellence Framework

In 1987 the Malcolm Baldrige National Quality Award was created, with the goal of increasing the competitiveness of U.S. organizations by focusing on quality. The scope of the awards has since expanded to organization-wide excellence [18]. The award focuses on performance and outcomes in five key areas; product and process, customer, workforce, leadership and governance, financial and market. To receive the award, an organization must meet the Criteria for Performance Excellence, which are a part of the Baldridge Excellence Framework, Fig. 16.1 [19].

The framework focuses on seven critical areas as displayed in the blue shapes on the grey platform in Fig. 16.1. It is based on 11 beliefs and behaviours which have been found in high performing organizations, shown in the blue base of Fig. 16.1 as core values and concepts. According to the framework, performance excellence consists of three things; improving value delivered to customers and stakeholders, improving overall effectiveness, and more knowledge for the organization and employees [2].

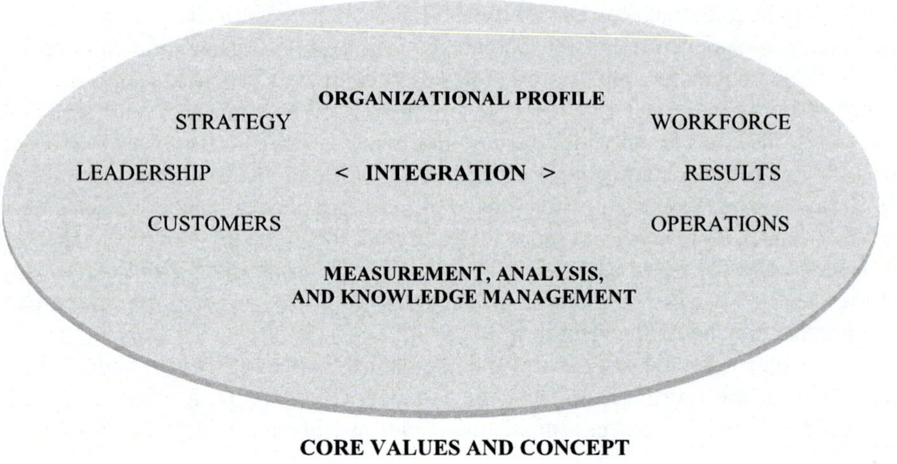

Fig. 16.1 The Baldridge excellence framework

16.2.2 EFQM Excellence Model

In 1988, to promote higher standards, leaders from fourteen companies founded the European Foundation for Quality Management, now the EFQM. In 1991, the EFQM Excellence Model was launched, and in 1992 the first EFQM Global Excellence Award was given. Since then, the model has been developed and extended [6]. The EFQM Excellence Model is based on nine different criteria as seen in Fig. 16.2, as well as on the Fundamental Concepts of Excellence, which are eight principles [7].

The framework is supposed to help organizations understand the relationship between what they do and what they achieve and assumes that an excellent organization will comply with the ten principles of the UN Global Compact [7]. According to the Excellence Model, "Excellent Organisations achieve and sustain outstanding levels of performance that meet or exceed the expectations of all their stakeholders" [7].

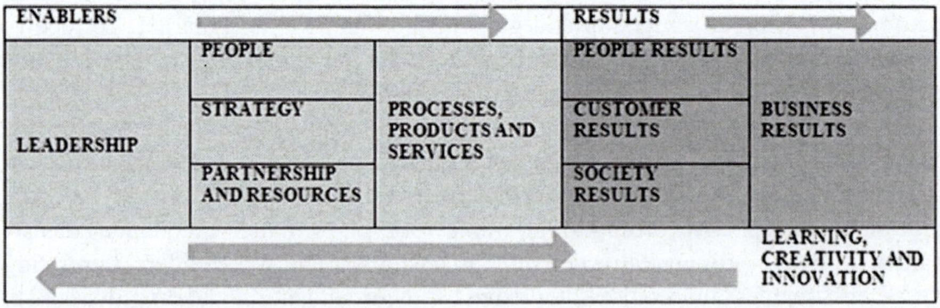

Fig. 16.2 EFQM excellence model

16.3 Excellence Canada

The National Quality Institution, now Excellence Canada, was established in 1992 by Industry Canada. The goal was to help organizations perform better and become more globally competitive by providing best practices [9]. The framework has since been expanded, and a more recent development is the Excellence, Innovation and Wellness Standard. The standard is split into five drivers; leadership, processes, people, customers and planning (Fig. 16.3) and claims to ensure that "organisations achieve the best results possible across all areas" [8].

The standard has a four-level staged approach to certification and organizations with a certification can apply for the Canada Award for Excellence. The definition of excellence is "meeting and exceeding rigorous standards and requirements, demonstration of continuous improvement, measurement of progress and verification" across all five drivers [10].

16.3.1 The Shingo Model

The Shingo Model was created as a part of Utah State University in 1988. The programme's goal is to recognize the best in enterprise excellence throughout the world. When using the Shingo Model, an organization's culture is aligned with the Shingo Guiding Principles [24] in Fig. 16.4.

The framework revolves around the three insights of enterprise excellence. All three insights are about how to influence culture and people's behaviour towards

Fig. 16.3 The drivers of excellence

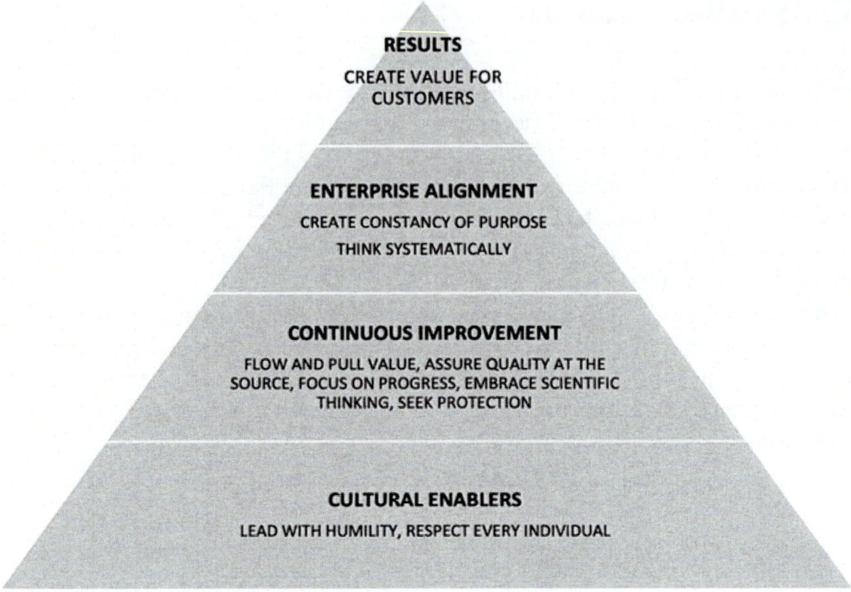

Fig. 16.4 Shingo guiding principles

the Shingo Guiding Principles. The first insight is about how that ideal behaviour will get ideal results. The second insight says that behaviour is driven by purpose and systems. The third insight states that if principles are behavioural rules that have consequences, then the better people understand the principles, the better they understand ideal behaviour. "Operation excellence requires ideal behaviour that translates into consistent and ideal results" [24].

16.3.2 Intersections

If the basic ideas of each of the models are compared, a certain trend begins to appear. The main concepts can be sorted into six dimensions; Leadership, Vision and strategy, Continuous improvement, Processes, People, and finally Universal success and Systematic approach. According to the models, excellent management of these six dimensions will lead to excellent performance when it comes to meeting or exceeding the needs and expectations of the four key stakeholder groups, the clients, the employees, the owners and the community [1]. In Table 16.1, the intersections of the models are shown by sorting their key concepts into the six dimensions.

When we compare how the models define excellence, another trend appears. The key drivers of success can be grouped into fulfilling or exceeding the needs of four key stakeholder groups; the customers, the employees, the owners and society [1].

Table 16.1 Excellence models and their intersections

Dimensions	Principles of business excellence models			
	Baldride	EFQM	Canada	Shingo
Leadership	Visionary leadership. Ethics and transparency	Leading with vision, inspiration & integrity	Leadership through culture, values and direction for success	Lead with humility
Vision and strategy	Organizational learning and agility. Focus on success	Managing with agility. Developing organizational capability	Developing strategic, business and improvement plans	Seek perfection. Create constancy of purpose
Continuous improvement	Managing for innovation. Management by fact	Harnessing creativity & innovation	Governance and innovation. Monitoring, evaluating & reporting on progress on strategic, business & improvement goals	Embrace scientific thinking
Processes	Customer-focused excellence	Adding Value for Customers	Prevention-based, rather than correction based. Management of supplier relationships	Assure quality at the source. Flow & pull value. Create value for the customer
People	Valuing people	Succeeding through the talent of people	Safety and wellness of employees and their families, physical and psychological dimensions	Respect every individual
Universal success and systematic approach	Systems perspective. Societal contributions. Delivering value and results	Creating a sustainable future. Sustaining outstanding results	Fulfil the organization's legal, ethical, financial and societal obligations	Focus on process. Think systematically

In short, according to the business excellence models, an organization cannot be excellent if it leaves out one of the four groups.

16.3.3 Corporate Social Responsibility

Festa is the Icelandic Center for Corporate Social Responsibility. It is a member-based not-for-profit association of organizations in Iceland. It is "dedicated to raising awareness of CSR and sustainability among Icelandic businesses and the general public, as well as supporting its member organizations in implementing sustainability and CSR into their strategy and day-to-day operations." [11]. According to FESTA, "In its purest form, corporate social responsibility (CSR) is about companies taking responsibility for the impact they have on people and the environment" [12]. Socially responsible companies organize their operations in a way that will not impact society or the environment in a negative way. There are many ways for an organization to practise CSR, and here I will name a few. An initiative named The Global Compact was established in July 2000 by the United Nations. Its mission is to "mobilize a global movement of sustainable companies and stakeholders to create the world we want" [27]. The compact is based on ten principles, which cover human rights, labour, environment and anti-corruption [28, 29]. In 2015, 17 highly ambitious goals were set in order to make the world a better place by 2030—The Global Sustainable Development Goals of the United Nations. The agreement involves sustainable developments on social, economic and environmental scales [13].

The International Organization for Standardization (ISO) brings experts together in order to share knowledge and develop standards with the aim of ensuring quality, safety and efficiency. It is an independent, non-government international organization [14]. One of those standards is ISO 26000: Guidance on social responsibility. It helps organizations be more socially responsible. It helps organizations to create effective actions and shares/share best practices [16]. The ISO 14001: Environmental Management Systems is grounded in mapping out a framework that establishes an effective environmental management system. A company with an ISO 14001 certification can measure and improve its environmental impact [15]. ISO 45001: Occupational Health and Safety Management Systems provides a framework a company can use to improve the occupational health and safety (OH&S) in the workplace [17]. IST 85 is the Equal Wage Management System. According to Icelandic laws, companies with 25 employees or more must be certified according to this standard to show that they are respecting the equal wage act [23]. Every company that has 25 or more employees must fulfil this standard, but the time limit for certification varies, depending on the company size. The range is from 31st December, 2019 to 31st December, 2022, with the smallest companies having the longest time [23].

16.4 Research Method

The objective of this research is to determine to what extent companies that have been deemed as excellent in Iceland are socially responsible. The authors try to achieve this by determining which companies are excellent according to the business excellence standards and measuring how socially responsible they might be deemed to be.

16.4.1 The Excellent Companies

In this research, excellent companies were determined/identified as all companies that appeared on Credit Info's Outstanding Companies list and VR's Exemplary Companies list in 2018. The lists of the 2018 winners were cross-referenced to find the companies which were on both lists. In total, the companies were 25. They ranged from small to medium to large, small having less than 20 employees and large more than 50. The size distribution between groups was almost completely even, with nine small companies, eight medium-sized and eight large ones. E-mail addresses of managers and directors were then found on company websites or by calling the companies.

16.4.2 The Survey

Originally, the authors wanted to interview all the companies. However, once they reached the final number of 25 companies, this was not deemed to be feasible for a paper of this scope. Then the choice was between getting a small glimpse of many of the companies or choosing three or five and getting detailed information from them. The benefits of doing a survey are that it offers an easy way to reach all the companies, which in turn, if the response ratio is sufficient, returns a broader spectrum of answers. Therefore, the authors resorted to doing a survey among these 25 companies.

The survey was sent to the CEOs of each of the "excellent" companies, followed by a reminder a week later, and a phone call in the course of the last days of the survey, so as to get a better response ratio. The survey was open from 16 April, 2019 to 3 May, 2019. It comprised nine questions, seven of which had yes or no answers and two of which were multiple-choice questions. A copy of the survey can be requested from the authors. In the end, CEOs of 19 companies answered the survey, making the response ratio 76%. It was a good mix of small to large companies, 8 small, 5 medium and 6 large corporations.

The survey questions addressed each company's social responsibility to four different groups of stakeholder/interested parties in the business excellence models; customers, employees, owners and society. As the authors started to assemble the survey, they decided to focus more on society and "making the world a better place", rather than on the other three groups. The reasons for leaving out the employees was that the companies had already been given the Exemplary Award that covers that part extensively. Furthermore, since the response was based on a self-reflection, the authors did not think it would give an accurate depiction of the current status on behalf of the customer group. The same applies to the shareholders (owners)—that is, they were not asked.

Quite a few articles have been written about how to measure CSR. However, none of the methods claim to be the best one. As a result, the survey questions were taken from a few different directions, some from the business excellence models, while others were from articles written on the subject. The authors tried to have a broad range of questions; global impact, the environment [26], standards, social initiatives [22] and finally whether CSR was incorporated into the company strategy and vision.

16.4.3 Limitations

When reading the survey and the interpretation of results, it should be borne in mind that the authors are approaching the subject out of interest rather than expertise. Also, there are a few limitations that must be addressed: (1) Firstly, an assumption is made that the awards chosen as a starting point are close enough to the corresponding dimensions in the business excellence models to be real indicators of excellence. (2) A second limitation is the companies themselves and the fact that they vary greatly in size, ranging from less than 20 to more than 100 employees. Also, the Exemplary Award survey is sent to all employees of the companies it evaluates. However, not all employees of every company answered the Exemplary survey done in 2018, and the response ratio ranged from low, 35–49%, to high, 80–100%. However, in the case of the survey conducted for the purposes of this paper, 14 of the 19 companies had a response rate of 80–100% in the Exemplary survey, with only one company in the 35–49% interval/range. (3) Finally, the measurement of CSR is not straightforward. A lot has been written on the subject, but each method has its limitations. A survey this size can never take into account all of the company's stakeholders, e.g. employees, customers, society, government, competitors, environment, future generations and nongovernmental organizations [26]. Also, since 11 of the 19 companies answered the survey independently online, the respondents may not have understood the questions in the way that the authors meant them to be understood. Therefore, the survey can only ever offer a glimpse of the real situation.

16.5 Research Results

The results are based on the answers of 19 companies to the nine questions in the survey. Eleven companies answered through an online survey, the other 14 companies were called up and asked to answer the survey via telephone with eight of them eventually reached and/or willing to answer. The questions were then sorted into the two different groups, local community and making the world a better place.

Seven questions were assigned to the local community group. The questions were: (1) Does the company offer jobs for disabled people? (2) Does the company support causes concerning people with disabilities? (3) Does the company offer different methods of practicing sustainability? (4) Does the strategy or vision of the company involve CSR? (5) Does the company fulfilled three different standards? (6) Does the company strategy and or vision involve social responsibility? (7) Does the company strategy and or vision involve social responsibility?

It turns out that most of the companies, i.e. 15 of them, support some causes concerning disabled people, while six offer jobs specified for them, see Fig. 16.5. During a phone call, one company mentioned that it offered jobs for people trying to get back to work after burnout or a long-term illness, and another said that they did not dismiss people of retirement age who wanted to keep working, the oldest person still working being 75 years old.

Almost all of the companies, or 16, practise corporate social philanthropy and many, 13, are also socially responsible in their business practices, Fig. 16.6. On the phone, one of the companies also mentioned that they regularly donate items to those in need.

A total of 15 answered that either or both the vision and strategy involved social responsibility, Fig. 16.7, and 11 of those answered yes to both. Only four out of the

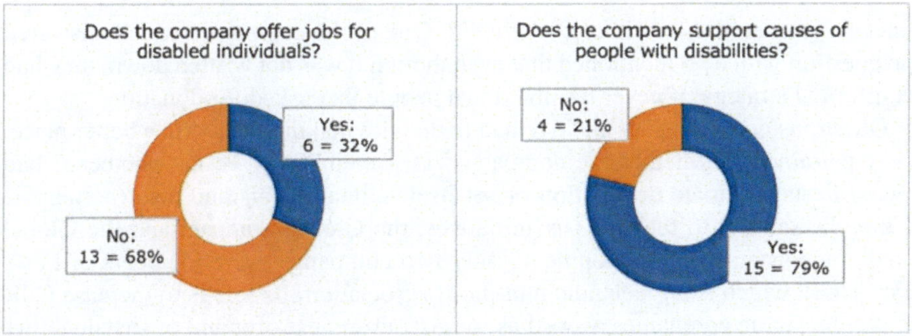

Fig. 16.5 Answers to Q8 and Q9

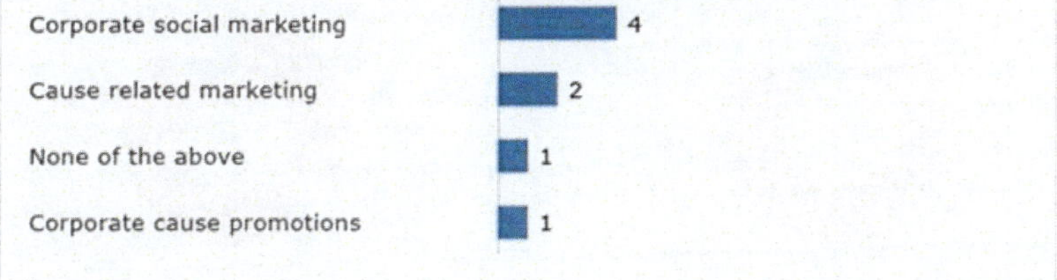

Fig. 16.6 Answers to Q7

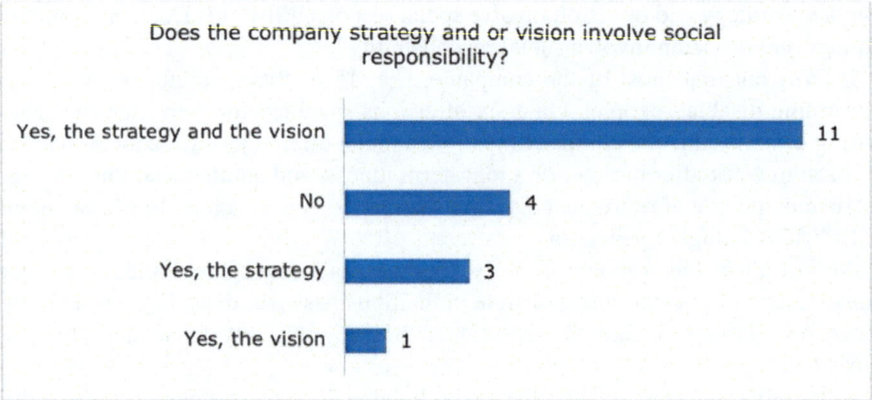

Fig. 16.7 Answers to Q6

nineteen answered that neither of them did. One of the companies which answered the question with a no mentioned that even though it was not written down, they had an informal strategy of never turning down people that asked for donations.

Questions one, two, four and five had to do with making the world a better place. They pertained to whether the company was a member of Festa, whether it had signed Festa's climate declaration or set itself climate goals and also (pertained? is this the sense?) to the two UN initiatives, the Global Compact and the Global Goals for Sustainable Development. Only two companies were members of Festa (Fig. 16.8), which is an Icelandic non-profit association. Its role is to increase CSR among Icelandic companies as well as to raise awareness of it and of sustainability among the general public. It also supports its members in the implementation of CSR and sustainability [11]. The answer to the climate goal question was an overwhelming 17 noes against 2 yeses. However, in the phone calls made to companies, two answered that even though they did not have a climate goal at the moment, they were in the middle of creating it. Another company mentioned that their eventual

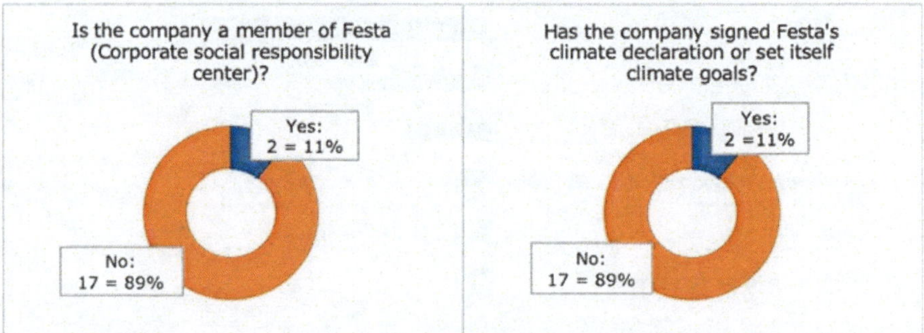

Fig. 16.8 Answers to Q1 and Q2

goal was to be a leader in environmental issues. Another company mentioned that even though they did not have climate goals written down, their product was very environmentally friendly, which was a primary goal of theirs.

Most of the companies had not signed the Global Compact nor chosen a goal from the Global Goals initiative (Fig. 16.9), with 16 noes against 3 yeses. One company was very interested in these initiatives and asked the authors to send over some information about them. The same three companies that have signed the Global Compact have chosen SDGs. One of them is also a member of Festa and has signed its climate declaration.

Question three was a multiple answer question, which touched on a few different standards and both stakeholder groups. For the local community group, there was ISO 9001, which is a quality management standard, ISO 45001 for health and safety and ÍST 85 for equal pay. For making the world a better place, there were two standards; ISO 26000, which provides guidance on how to operate globally in a socially responsible way and ISO 14001, which is an environmental standard. Out of the 19 companies that answered, 13 did not fulfil any of the standards (Fig. 16.10). When phoned, two of the companies said they had never heard of some of them and

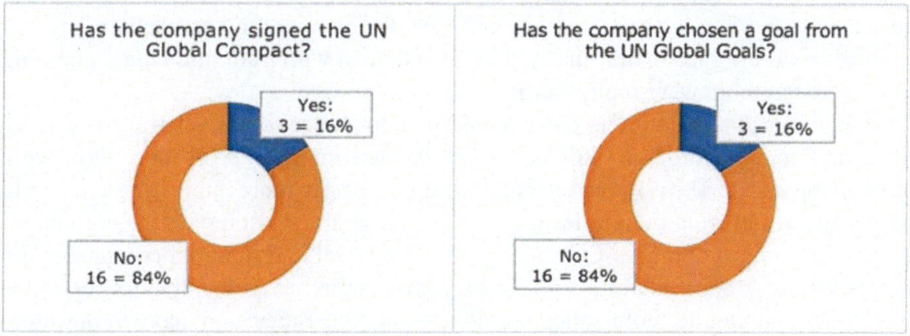

Fig. 16.9 Answers to Q4 and Q5

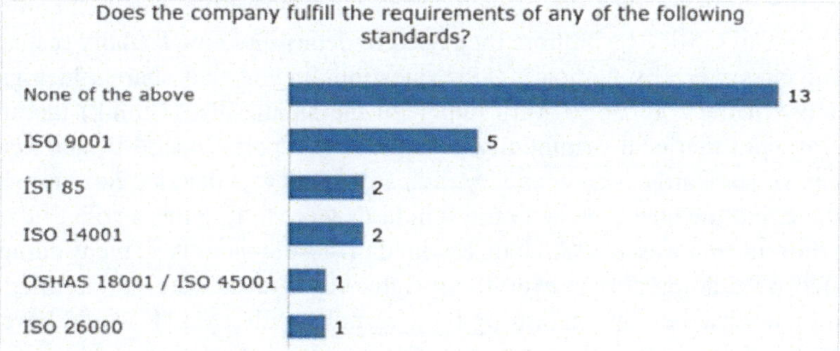

Fig. 16.10 Answers to Q3

asked the authors to send them more information. One company mentioned that they were working towards fulfilling ÍST 85 very soon, and two others mentioned that they would not fulfil it because of their size. Another two companies said that even though they themselves did not fulfil these standards, all their manufacturers did, and a third company said that they were working towards being certified in ISO 45001 and fulfilled the other ones but were certified by a third party.

When individual company responses are compared, the companies vary, but with two extremes. One company answers all questions but one in a negative way while the other one answers all questions, except one, in a positive way. All the other companies are somewhere in the middle.

16.6 Discussion

After all of the companies within the survey had been contacted by telephone ?? (via a phone call to those eight companies omit words in this bracket ??), the authors realized that it would have yielded more accurate results to call all the companies and ask the questions herself. The discussions that took place during the phone calls gave the authors a much better insight into what the companies were really doing and showed her a few gaps in her survey. This also helped with a common understanding of what the authors were really asking.

One gap is that the authors did not ask the companies to answer if the company was small, medium or large. This was a significant oversight, and as the authors went through the results, she discovered that it would have been more interesting and would probably have given a more relevant outcome to be able to compare the companies within the same size range. Comparing how the small and medium companies did versus the large ones would also have been interesting, as in the authors' opinions it is likely that smaller companies are doing less, since they may have to put more focus on day-to-day operations than on CSR. It is impossible to corroborate with this dataset, but the authors' opinions were strengthened by the phone calls made to the smaller companies.

From the survey results, it is reasonable to assume that the excellent companies are all aware of CSR. The authors then tried to determine how socially responsible the companies were by comparing the questions in the two shareholder groups. Since most of the companies were rather on the small side, (I think) the authors decided to split the local community group into two parts, A and B, and keep the fulfilment of standards as a secondary indicator. This was decided because she did not believe that the questions with the standards were telling the whole truth. Even though the question was not, "Are you certified in these standards?", many companies she called were doing almost everything right, but as they were not certified, they were not comfortable with answering the questions with a yes. It would have been better to ask the companies questions that were less leading such as "Do you have an operational quality management system?".

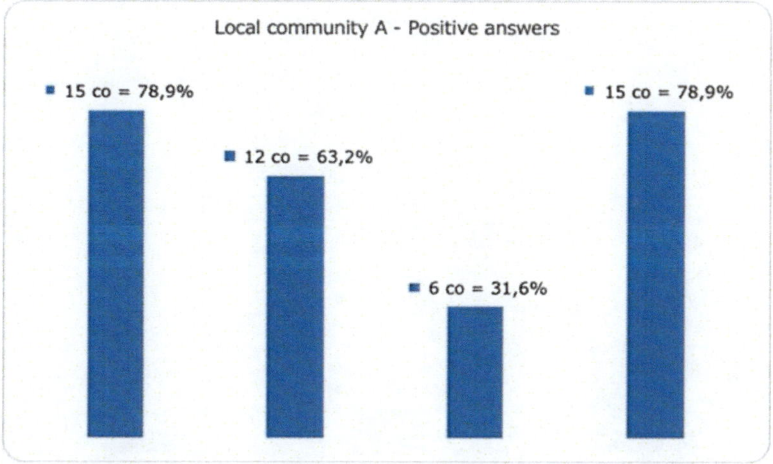

Fig. 16.11 Local community A—Percentage of companies, co, with positive answers

The authors were pleasantly surprised by how well the companies did on part A on local community (Fig. 16.11). All questions got a positive answer percentage of over 60%, except the question about work for disabled people. In the question about different types of CSR practised, two or more types of CRS practised were required for a positive result.

A gap in the survey was made clear when one company talked about how they assisted people who were trying to get back to work after, e.g. illness or burnout. This was not addressed in the survey but could have been rectified by having an open question at the end of the survey about what else a company was doing. However, since this came up in a discussion it is not certain that such a solution would have been adequate.

The companies did not do as well in part B. It was disappointing to the authors to see that only one company fulfilled either the ISO 45001 or the OSHAS 18001 standard and just five companies fulfilled ISO 9001. ÍST 85 is a different matter. By law, a company does not have to get certified if it has less than 25 employees, and the first companies have a deadline until 31st December, 2019. Since eight out of the 19 answers were from companies with fewer than 20 employees and the survey was done in the first quarter of 2019, it is not a big surprise to see that only two of the companies fulfil ÍST 85. However, one company mentioned that they were working towards getting certified as soon as possible [25].

In the dimension of making the world a better place, the results were not very positive (Fig. 16.12). However, it was encouraging that one company was very interested in the UN Global Goals and asked the authors to send some information about them, and another wanted more information about the standards. There is definite room for improvement in this dimension, and the authors believe it shows that the companies are more concerned about the local community they operate in than about the rest of the world.

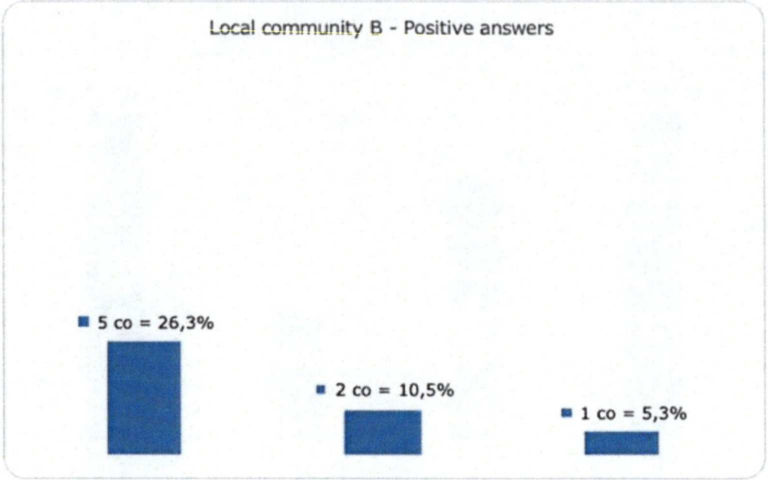

Fig. 16.12 Local community B—Percentage of companies, co, with positive answers

The results show many good examples of CSR on a local community scale, but in the end the question is how the companies are doing over both dimensions. It is not possible to say that as a group these companies are socially responsible. On the global scale, taking the first step has been made easy by organizations such as Festa and the UN, and anyone can choose a global goal suited to their operations (Fig. 16.13).

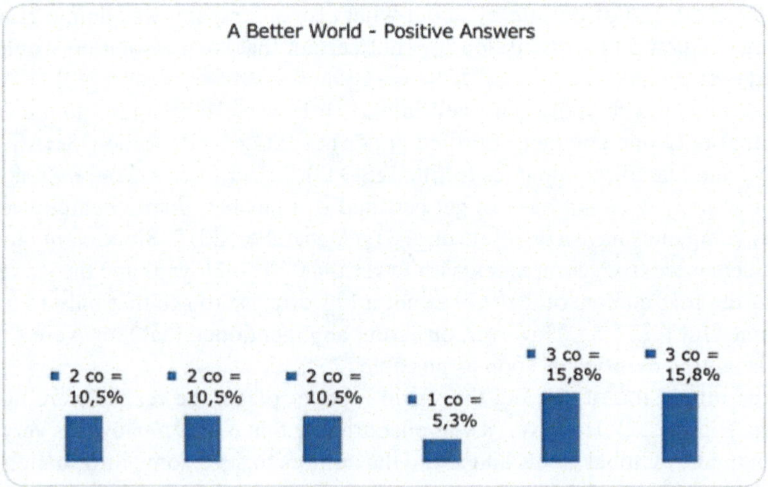

Fig. 16.13 A Better World—Percentage of companies, co, with positive answers

Incidentally, to what standard do we hold companies with awards for not only being outstanding but exemplary as well? The premise of this paper is that these companies must be the best. However, the international business excellence models all agree that a company cannot be truly excellent without fulfilling the needs of society and customers as well as owners and employees. The authors fully support the acknowledgment of companies for a job well done, but these award titles are misleading and need to be called something more descriptive of what is really being measured.

16.7 Conclusion

With the interest in CSR on the rise [5] and the world becoming smaller and smaller as technology advances, the importance of CSR keeps growing. The result of this survey shows that Icelandic excellent companies are more invested in their local community than in the world at large. Of course, everyone must start somewhere, and it looks as if they are off to a good start. However, it must be said that with recent threats to the world we live in, such as Global Warming [20] and an ever increasing plastic pollution [21], thinking locally may not be enough, and it would be a shame if Icelandic excellent companies were to ignore their role in global society.

Compliance with Ethical Standards

We declare that we ensured the objectivity and transparency in our research and that accepted principles of ethical and professional conduct have been followed. Prior informed consent was obtained from individual participants included in the study before the research. No sensitive personal data was accessed. Anonymity of individual participant data is maintained. The research does not require ethics approval, as it mentioned in the waiver no. 11-57-59, issued by the Reykjavík University (RU) Ethics Committee.

References

1. Gunnarsdóttir AH, Ingason HT (2017) Afburðastjórnun. JPV útgáfa, Reykjavík
2. Baldige Program: *Baldrige Excellence Builder*. (2019) Retrieved May 3, 2019 from https://www.nist.gov/sites/default/files/documents/2019/02/06/2019-2020-baldrige-excellence-builder.pdf
3. Credit Info: Það er eftir eftirsóknarvert að skara framúr (ed) Retrieved May 3, 2019 from https://www.creditinfo.is/framurskarandi/default.aspx
4. Credit Info: Hvað gerir fyrirtæki Framúrskarandi?. (2018, 24. August) Retrieved May 3, 2019 from https://www.blogg.creditinfo.is/blog/hva%C3%B0-gerir-fyrirt%C3%A6ki-fram%C3%BArskarandi
5. Cone: 2017 Cone Communications *CRS Study*. Boston (2017)
6. EFQM. What is the history of EFQM? (ed) Retrieved May 3, 2019 from http://www.efqm.org/index.php/knowledge-base/what-is-the-history-of-efqm/

7. EFQM. *EFQM Excellence Model 2013*. Brussels: (2012)
8. Canada E (2017) Excellence. Innovation and Wellness Standard, Toronto
9. Excellence Canada: Our History (e.d.-a) Retrieved May 3, 2019 from https://excellence.ca/about-us/#H
10. Excellence Canada. The Standard of Excellence (ed-b). Retrieved May 3, 2019 from https://excellence.ca/canada-awards-for-excellence/
11. Festa: Framtíðarsýn festu (2015a) Retrieved May 3, 2019 from https://festasamfelagsabyrgd.is/#lead1
12. Festa: Hvað er samfélagsábyrgð? (2015b) Retrieved May 3, 2019 from https://festasamfelagsabyrgd.is/
13. The Global Goals: The 17 goals. (ed) Retrieved May 3, 2019 from https://www.globalgoals.org/
14. ISO: About us. (ed-a). Retrieved May 3, 2019 from https://www.iso.org/about-us.html
15. ISO. ISO 14000 family—Environmental management (ed-b) Retrieved May 3, 2019 from https://www.iso.org/iso-14001-environmental-management.html
16. ISO: ISO 26000—Social responsibility (e.d.-c). Retrieved May 3, 2019 from https://www.iso.org/iso-26000-social-responsibility.html
17. ISO: *Occupational health and safety—ISO 45001*. Geneva (2018)
18. NIST: History (ed-a) Retrieved May 3, 2019 from https://www.nist.gov/baldrige/how-baldrige-works/about-baldrige/history
19. NIST: Improve Performance and Get World-Class Results. (ed-b) Retrieved May 3, 2019 from https://www.nist.gov/baldrige/baldrige-award
20. Nunez C (2019) Causes and effects of climate change. (2019, 22. January) Retrieved from https://www.nationalgeographic.com/environment/global-warming/global-warming-overview/
21. Ostle C et al (2019) The Rise in Ocean Plastics Evidenced from a 60-Year Time Series. Nature Communications 10(1):1622
22. Sirgy MJ, Lee DJ (2008) Well-being marketing: an ethical business philosophy for consumer good firms. J Bus Ethics 2008(77):377–403. https://doi.org/10.1007/s10551-007-9363-y
23. Staðlaráð Íslands: Jafnlaunastaðallinn ÍST 85—Um hvað er hann og til hvers? (2017, 13. June) Retrieved May 3, 2019 from https://www.stadlar.is/thjonusta/nyjustu-frettir/stadlamal-frettabref-stadlarads/2017/06/jafnlaunastadallinn-ist-85-um-hvad-er-hann-og-til-hvers.aspx
24. Shingo Institute: https://shingo.org/
25. Stjórnaráð Íslands: Algengar spurningar (e.d.) Retrieved May 3, 2019 from https://www.stjornarradid.is/verkefni/mannrettindi-og-jafnretti/jafnretti/jafnlaunavottun/algengar-spurningar/
26. Turker D (2008) Measuring corporate social responsibility: a scale development study. J Bus Ethics 2009(85):411–427. https://doi.org/10.1007/s10551-008-9780-6
27. United Nations: Our Mission (ed-a). Retrieved May 3, 2019 from https://www.unglobalcompact.org/what-is-gc/mission
28. United Nations: The Ten Principles of the UN Global Compact (ed-b) Retrieved May 3, 2019 from https://www.unglobalcompact.org/what-is-gc/mission/principles
29. United Nations: What is the commitment (ed-c) Retrieved May 3, 2019 from https://www.unglobalcompact.org/participation/join/commitment
30. VR: Lykilþættirnir níu (ed) Retrieved May 3, 2019 from https://www.vr.is/kannanir/fyrirtaeki-arsins-2018/lykilthaettirnir-niu/

Part III
Perspectives of Integrating Sustainability into Project Management: Learning and Continuing Competences Development

Chapter 17
Chatbot-Based Training for Project Management: Another Way of Corporate Training or a Must-Have Tool for Sustainable Education?

Constanta-Nicoleta Bodea, Maria-Iuliana Dascalu, and Alexandru Hang

Abstract Chatbots or virtual conversational agents are computer programs that use natural language processing techniques to understand and communicate with people via a chat interface, which is usually textual or auditory. There are a lot of successful use cases of chatbots nowadays, having a wide area of applications, from e-commerce, customer relationship management, media, healthcare, project management and education. We claim that chatbots are very useful in corporate training, as they support professional development and knowledge sharing in day-to-day business practices, they reinforce learning and help professionals to find quickly the required information, thus contributing to a sustainable education. In this paper, we propose a pattern-based chatbot, useful for corporate training for competence enhancement. The solution was successfully deployed in a professional association and integrated in its digital platform.

Keywords Project management · Training · Professional development · Chatbot · Sustainable education

17.1 Introduction

Chatbots are computer programs artifacts, not bound to any specific device and created for continuous user interactions. Their main purpose is to answer quickly to users' questions, based on a query to a large data storage. Bots are assigned a certain degree of artificial intelligence and, with continuous training, their efficiency

C.-N. Bodea (✉)
Bucharest University of Economic Studies, Bucharest, Romania
e-mail: bodea@ase.ro

M.-I. Dascalu · A. Hang
University POLITEHNICA of Bucharest, Bucharest, Romania
e-mail: maria.dascalu@upb.ro

A. Hang
e-mail: alexandruhang@yahoo.com

R. Cuevas et al. (eds.), *Research on Project, Programme and Portfolio Management*, Lecture Notes in Management and Industrial Engineering,
https://doi.org/10.1007/978-3-030-60139-3_17

is expected to increase [23]. Generally, chatbots do simple tasks in a repeated manner, but much quicker than a human agent could, that is why they are used in more and more fields nowadays, from business to education [28, 29]. At the base of the first conversational agents the Turing test lays as a criterion of intelligence. This criterion describes the ability of a computer to act as a human being in a real-time typed conversation with a real human and to be impossible to distinguish between the computer program and the real person [12]. In the present time, there are two types of chatbots, one that use patterns or rules and another one that is more advanced and use machine learning technology. Both have their own advantages and special purpose. In the case of chatbots based on patterning or rules, they are limited in matters of answering the subject, as they reply only to specific patterns and don't go deeper to determine the meaning, sentiment or intent of the input. But a beneficial characteristic of this type of bots is that the developer controls the degree of intelligence of the bot and what questions to answer to. These bots are oriented on specific domains and the interaction needs to follow a certain workflow. Chatbots based on artificial intelligence not only recognize patterns, but learn repetitively and are able, after training, to understand the meaning, intent and sentiment of what the user types. The main advantage of this type of bots is that they get smarter in time. They are used in advanced, multi-purpose scenarios.

We claim that chatbots are a valuable instrument for sustainable education, as they offer long-lasting solutions for active learning, related to social, economic and environmental issues, which are the three pillars of sustainability [2]. They can be used for lifelong learning, while people are engaged in working activities, without spending extra money or extra time on updating one's knowledge to a specific concept [8, 21].

There are lot of chatbot APIs which allow the fast building of a conversational bot, e.g. Wit.ai, Google DialogFlow (former Api.ai), Luis.ai [18]. Chatbot APIs are in fact natural language processing (NLP) services. Using a knowledge base and machine learning algorithms, these systems can learn to extract data and meaning from sentences by exercising on various inputs. Each API has its own drawback and, for a specific domain, the chatbot created with a certain API needs a lot of training.

In this paper we propose a domain-specific chatbot, therefore we decided to implement it without a well-known chatbot API, but by using patterns. Our PM-CERT chatbot is useful for project management training. It was deployed by the Project Management Romania professional association, which is affiliated to International Project Management Association (IPMA). Our work is presented in the framework of other successful chatbots' use cases.

17.2 Chatbots' Use Cases

There are a lot of successful use cases of chatbots nowadays, having a wide area of applications. A common utility of chatbots is to provide services to the companies' customers by using a familiar communication media, such as companies' websites or available social networks, and are more engaging than mobile apps [6].

In e-commerce, bots can help people order almost anything, provide useful information and even take care of the payment in some cases [27]. For example, instead of going through a cinema's website, searching and buying a movie ticket, a customer can ask about the most viewed movies of the week and the bot, using the knowledge base and the cinema's database, can make recommendations to the user, can show him trailers and promoting media and lastly, proceed to make the reservation or buy the ticket. The advantage is that everything is done from a sole source with the use of human readable messages. The customer is not anymore needed to check various websites, work with several, different interfaces to decide. Everything can be achieved from the chat box, saving valuable time.

Customer relationship management is also a promising field of chatbots, as they can provide 24/7 support regarding the company's products and services, they can manage the post sales and follow-up processes and speed up the interaction between the user and the firm. Rather than contacting a customer representative, the user will receive its service and answers almost immediately from an automated system, without the need to wait in a queue. It's clear that customer relationship is improved, taken to a new level, where information is accessed much faster and presented clearly when desired. Gartner, an American research and advisory company, claims that human interaction will be needed only in one third of the customer-service interaction process [22].

Chatbots are also used by businesses in the media sectors as content delivery services [6]. The Wall Street Journal, Tech Crunch, CNN and other publications use chatbots to deliver relevant news to their customers. Interacting with the interface, the user selects several domains he's interested in and the bot will send him articles when they are available, removing the need to browse their websites and search for it. After 6 months of chatbot experimentation, CNN representatives said that "the most effective chatbots brought the right mix of editorial curation and technological automation" [19]; so they admitted that chatbots cannot solve the dilemma of distribution by themselves, but they can reach "every individual with a unique message at a unique moment in time" [19], fact that cannot be done by a human editor. So, chatbots are enablers of scale and granularity regarding information dissemination and provide timely customization of it. Conference bots are used to schedule meetings and activities for an event, offer information about participants and speakers and act as mediators between members [6]. Lastly, they can gather feedback from the participants in any moment and presenting the results, removing the need to ask each one of them and analyze yourself the answers.

Healthcare solutions also started emerging [9]. Pocket coaches and nutritionists make suggestions on how one can improve his lifestyle, offering feedback based on what one should eat, drink, sleep and exercise. It also tracks one's daily activity and makes recommendations using a database generated by health experts and nutritionists around the world. There are also applications that give medical advice and possible treatments for different symptoms and illnesses, and although free, they help boost a clinic's image.

There are a few successful deployments of chatbots for the implementation of various project management processes in small business as well, e.g. in project collaboration, task management, usability and integration [30], in keeping the data consistency regarding the timesheets, risk prediction [5]. The Meekan chatbot [20] provides team members support to automatically match schedules, thus improving collaboration: instead of a long chain of e-mails, the chabot handles the meeting schedules within the project. The PMbot is able to automatically create status reports, avoiding useless face-to-face meetings [30]. Ayoga ActBot sends timely alerts to project members reminding them to fill timesheets or update their work progress through a friendly user interface [4].

17.3 Chatbots in Sustainable Education and Training

In the field of sustainable education, chatbots are seen as virtual assistant for various activities, from administrative tasks to tutoring and training ones, e.g. the admission process and students' orientation can be done by a chatbot: AdmitHub platform [1] which created educational chatbots used by Georgia State University, University of Memphis, West Texas A&M and Arizona State University is a good example in this respect. The chatbot Alex was developed by the Technical University of Berlin to help students find and schedule the courses more efficiently [3]. Jill Watson chatbot [13], based on IBM Watson platform, was developed by a professor of Artificial Intelligence for his class, in order to scale personalized learning, as he claims that the best learning experience for a student is when he establishes a personalized connection with the teacher. The chatbot had the capacity of increasing the students' engagement [17]. Other chatbot combines a well-known chatbot framework, Pandorabots [25] (PANDORA), with virtual reality technology, and creates an immersive training environment for successfully passing interviews in various domains [28, 29].

For the particular case of corporate training, chatbots seem to be a saving solution: corporations invest a lot of money into training their employees, with the purpose of their employees to obtain professional certification and/or to apply the trainings to day-to-day business practices [11]. Unfortunately, people tend to forget to recap the things they learnt during training, as they are more focused to immediate tasks.

Chatbots could automatically provide reminders, track goals, provide recap exercises for employees, reinforce learning, provide continuous performance [11] or help them search the proper content for preparing themselves for a certification or for getting their job done [24, 26]. Some successful use cases of chatbots in corporate training are described by Kowalski et al. [15]: 106 users and security specialists participated in a security training, in which they used an e-learning package, including a chatbot named Sally. The attitude of trainees who participated in the experiments was more positive when chatbots were used, compared to the traditional use of conventional e-learning for security training.

A lot of trainings in corporations are focused towards employees' preparation for a certification. Project management certification, which are usually very expensive, are among the certification paid by companies. Levinson [16] underlines the importance of certification nowadays, not only for individuals, but for firms, also: she cites an IBM CEO who says that clients request certified project managers to be part of the project team, as this is a guarantee of the project success.

Also, The Standish Group (https://www.standishgroup.com/), the publisher of the CHAOS reports which track IT project success and failure rates says that two-thirds of the interrogated persons regard project management certification as valuable [16]. Face-to-face trainings are costly, not only from money point of view, but from time investment point of view. A chatbot could be an instrument to provide customized help in preparation for getting a certification and using the timeframe suitable for each employee.

17.4 PM-CERT Chatbot

Taking into account the opportunities provided by the chatbot technologies nowadays and the number of potential users, we propose a chatbot-based training environment for project management. Our chatbot was deployed by Project Management Romania, a professional association which is affiliated to the International Association of Project Management (IPMA). The Romanian association offered it as a training instrument for project management professionals.

17.4.1 Description: Functionalities and Technologies

The goal of our chatbot was to help professionals to prepare themselves for a successful project management career. The knowledge repository or the database of the chatbot is based on the concepts included in IPMA standards—Individual Competence Baseline (ICB) [14].

The entire bot is based on the Natural Language Processing (NLP) of the user input, taken either from a website form (http://pmr-cert.ase.ro/ChatBot/) or from Facebook Messenger using Facebook Graph API [10]. The chatbot has a database of the most frequently asked questions by individuals, and some possible answers to them. In order to find the correct answer to send back to the user, it will search for some specific keywords in the user's input and try to find the most appropriate answer.

If no keyword is matched, then the Chatbot will send a default response, in which it says that the user should wait for a human to read the message. When a human agent answers, the chatbot learns from that answer. So, continuous learning is applied: if the chatbot doesn't know how to respond to a question, he will check what the page admin responded to the request and will memorize the response. The workflow of the chatbot is available in Fig. 17.1.

Some examples of questions, as well as the algorithm implemented in the main controller is depicted in Fig. 17.2. The chatbot can handle two types of user input:

- Facebook Messenger (through CURL Request to Graph API); CURL software is used in command lines or scripts to transfer data [7];

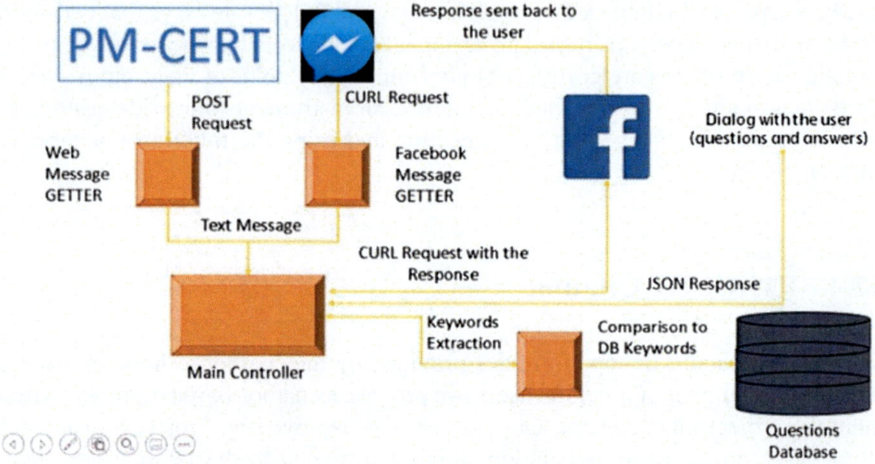

Fig. 17.1 Workflow of PM-CERT chatbot

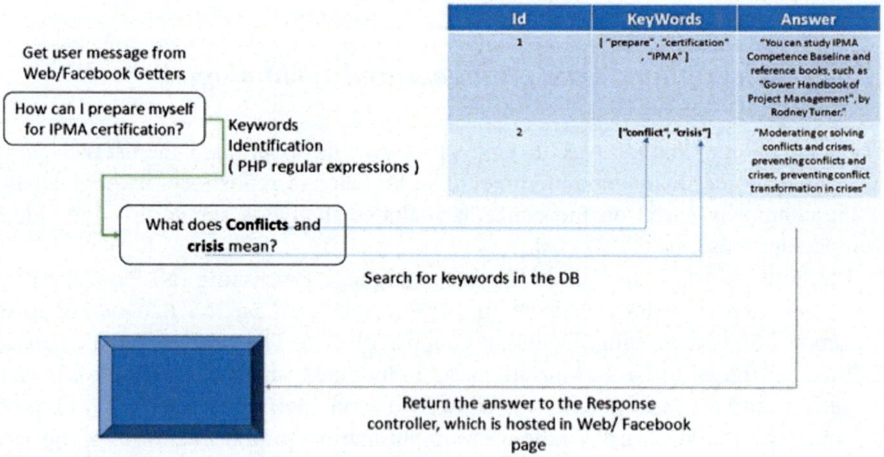

Fig. 17.2 Functionalities of chatbot main controller

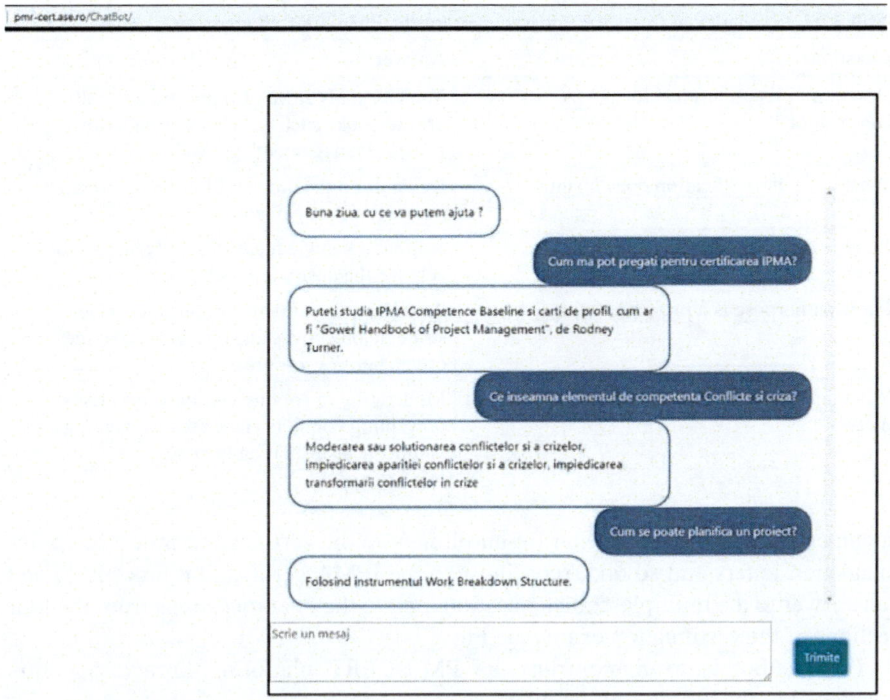

Fig. 17.3 PM-CERT chatbot visual aspect

- directly on the website http://pmr-cert.ase.ro/ChatBot/ (through Web POST Request).

The PM-CERT chatbot is web-based and was built using the following stack of technologies: PHP, SQL, Facebook Graph API, HTML, CSS, JS, Bootstrap, JQuery.

The chatbot itself is available in Fig. 17.3: it is a web-based form in which the user can insert text-based questions and receives text-based answers. For the moment, the chatbot is available just in Romanian, but examples of questions and answers translated in English can be seen in Table 17.1. Also, the keywords which are important to find a suitable answer are underlined in the text of each question.

17.4.2 Integration in the PMR Association Platform

The chatbot was created within the initiative of Project Management Romania to digitalize its processes and to save time of the association administrators as well as of the professionals. Project Management Romania offers a web platform (PM-CERT Platform) in which the ones who require certifications can upload the necessary

Table 17.1 Examples of (question, answer) pair for the PM-CERT chatbot

Question	Answer
How can I prepare myself for IPMA certification?	You can study IPMA Competence Baseline and profile books such as "Gower Handbook of Project Management" by Rodney Turner
What level of certification does it suit me?	Depends on the number of years of experience in project management
What is individual competence?	Applying knowledge, skills and abilities to achieve the desired results
For what purpose is a program defined?	A program is defined in order to achieve strategic objectives and optimal use of the organization's resources
What does conflicts and crisis as a competence mean?	Moderating or solving conflicts and crises, preventing conflicts and crises, preventing conflict transformation in crises

documents (e.g. CV, application for enrollment to the certification process, recommendation letters and so on, accordingly to the IPMA certification procedure) and can answer to the multiple-choice questions and to the open-questions from the four sections of the certification exam (see Fig. 17.4).

The chatbot is available from the PMR-CERT platform, through a button positioned in the left part of the interface.

The chatbot is not available anymore once the user starts simulating exams, which is visible in Fig. 17.5.

The first experiments showed that the exam simulation sessions are preferred by the platform users, but the interest for chatbot is gradually increasing, especially in the case of the new users.

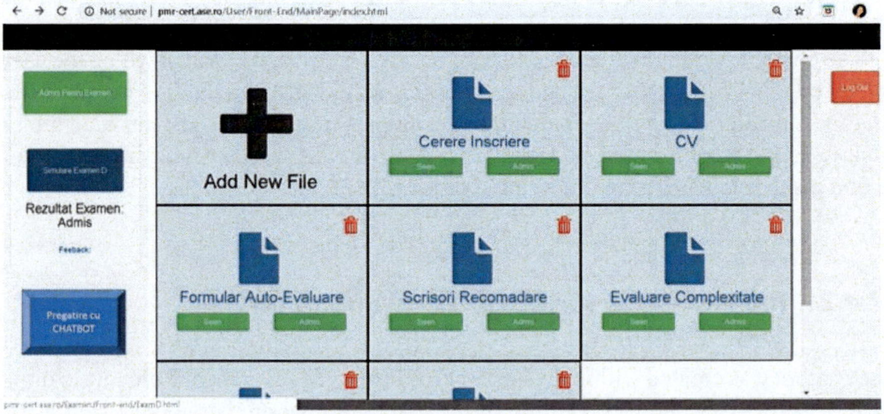

Fig. 17.4 Chatbot integration in the PM-CERT platform

Fig. 17.5 Simulation session using PM-CERT platform

17.5 Conclusions and Future Directions

In the future, more advanced NLP techniques will be used, based on machine learning or deep learning algorithms. Also, we plan to implement the processing of the input in multiple languages, like Microsoft LUIS API [18]. Although so far just a small group of users tested our chatbot (about 30 users who gave the certifications this year), all of them admitted the efficiency of the provided virtual teaching assistant, as they found the information quicker, without waiting for the human feedback. Consequently, as a follow-up of the technological boom, an education chatbot for corporate training and for preparation for professional certifications is no longer a fad or another way of modern training, but a necessity, a must-have tool for the modern society, in which sustainability of actions should be among main goals and objectives.

References

1. ADMITHUB (2019) an intelligent conversational platform for student success. https://www.admithub.com/. Accessed 8 May 2019
2. Ahamer G (2014) GISS and GISP facilitate higher education and cooperative learning design. In: Mukerji S, Tripathi P (eds) Handbook of research on transnational higher education. IGI Global, pp 1–21
3. ALEX Chatbot (2019) Alex. https://alex.qu.tu-berlin.de/. Accessed 14 May 2019
4. AYOGA (2019) Ayoga ActBOT. https://www.ayoga.in/construction-project-management-software-features#Ayoga-ActBOT. Accessed 8 May 2019
5. Biswas S (2018) The role of bots and AI in project management. https://chatbotslife.com/the-role-of-bots-and-ai-in-project-management-19d88f5aeca1. Accessed 7 Sept 2018
6. Cotrupe J (2018) Conversational A.I.: it's a bot time for a new conversation of customer engagement. Stratecast/Frost & Sullivan. www.nextit.com/downloads/SPIE-2016-15-Conversational-AI.pdf. Accessed 22 Apr 2018

7. Curl (2019) Command line tool and library for transferring data with URLs. https://curl.haxx.se/. Accessed 8 May 2019
8. Dascalu MI, Lazarou E, Constantin VF (2019) Technology model to support the initiation of innovation artefacts. In: Kravets AG, Groumpos PP, Shcherbakov M, Kultsova M (eds) Communications in computer and information science (CCIS), vol 1083. Springer, Cham, pp 278–287
9. Fernandes A (2019) The best chatbot apps in 2019—top 20 chatbot apps powered by AI. https://blog.verloop.io/the-best-chatbot-apps-2019-top-20-chatbot-apps-powered-by-ai/. Accessed 10 Jan 2019
10. Graph API (2019) Graph API description. https://developers.facebook.com/docs/graph-api/. Accessed 12 Jan 2019
11. Han V (2017) Are chatbots the future of training? https://www.shrm.org/resourcesandtools/hr-topics/technology/pages/are-chatbots-the-future-of-training.aspx. Accessed 10 Jan 2019
12. Hill J, Ford WR, Farreras IG (2015) Real conversations with artificial intelligence: a comparison between human–human online conversations and human–chatbot conversations. Comput Hum Behav 49:245–250
13. IBM Watson (2019) Watson. https://www.ibm.com/watson. Accessed 14 May 2019
14. ICB (2015) IPMA standards. https://www.ipma.world/individuals/standard/. Accessed 5 May 2019
15. Kowalski S, Pavlovska K, Goldstein M (2009) Two case studies in using chatbots for security training. In: Information assurance and security education and training (WISE 2009). Springer, Auckland, p 265
16. Levinson M (2010) Why project management certifications matter. https://www.cio.com/article/2421276/why-project-management-certifications-matter.html. Accessed 20 Jan 2019
17. Lipko H (2016) Meet Jill Watson: Georgia Tech's first AI teaching assistant. https://pe.gatech.edu/blog/meet-jill-watson-georgia-techs-first-ai-teaching-assistant. Accessed 10 Nov 2016
18. Luis (2019) Language understanding (LUIS). https://www.luis.ai/. Accessed 8 May 2019
19. McEleny C (2016) What CNN has learnt after six months of chatbot experimentation. https://www.thedrum.com/news/2016/11/16/what-cnn-has-learnt-after-six-months-chatbot-experimentation. Accessed 16 Nov 2019
20. MEEKAN. (2019) MEEKAN. https://meekan.com/slack/. Accessed 8 May 2019
21. Mogos RI, Bodea CN, Dascalu MI et al (2018) Technology enhanced learning for industry 4.0 engineering education. In: Revue roumaine des sciences techniques 63(4):429–435
22. Moore S (2017) 4 uses for chatbots in the enterprise. https://www.gartner.com/smarterwithgartner/4-uses-for-chatbots-in-the-enterprise/. Accessed 4 Feb 2019
23. Mrkalj M (2017) AI-powered chatbots to deliver better banking experiences. https://chatbotsmagazine.com/ai-powered-chatbots-to-deliver-better-banking-experiences-802f3c8ee75f. Accessed 3 Mar 2019
24. Oesch T (2018) Our "digital friends": using chatbots in corporate training. https://trainingindustry.com/articles/learning-technologies/our-digital-friends-using-chatbots-in-corporate-training/. Accessed 5 Apr 2019
25. Pandora (2019) Pandorabots. https://home.pandorabots.com/home.html. Accessed 14 May 2019
26. Popescu A, Dascalu MI, Bodea CN et al (2018) Towards a smarter recruitment process using chatbots. In: Paper presented at The 17th international conference on informatics in economy, INFOREC, Iasi, April 2018
27. Quoc M (2017) 10 ecommerce brands succeeding with chatbots. https://www.abetterlemonadestand.com/ecommerce-chatbots/. Accessed 6 April 2019
28. Stanica I, Moldoveanu F, Dascalu MI et al (2017) Considerations for a virtual training system to improve job interview skills for software engineers. In: Mihăescu C, Forbrig P (eds) RoCHI—international conference on human-computer interaction MATRIX ROM, Craiova, p 107

29. Stanica I, Dascalu MI, Bodea CN et al (2018) Virtual reality training system for improving interview performance. In: 2018 zooming innovation in consumer technologies conference (ZINC), IEEE Xplore, NoviSad
30. Tamang P (2019) 3 small business use cases of AI chatbots in project management. https://www.softwareadvice.com/resources/ai-chatbots-project-management-smb/. Accessed 6 Apr 2019

Chapter 18
Somebody Has to Teach Them, But What? Moving Towards Sustainable Teaching of Project Management

Steven A. Nijhuis

Abstract Teaching of project management is not without critique. Research in designing curricula for project management shows that the standard approach to competence research can easily produce the wrong competences. Comparing commercial and higher education offerings of project management shows that these target the same inexperienced group with the same approach to teaching, testing and contents. Critical competences show promise of defining the profession, but also show a high context dependability. There is a striking similarity between average junior and senior critical competences. Critical processes seem to be less context dependent. To get to sustainable teaching of project management, a focus on teaching processes is suggested.

Keywords Critical competences · Higher education · Teaching · Commercial education · Competence research

18.1 Introduction

Project management as a skill continues to grow in importance [36]. The gap between the employers' need for skilled project management workers and their availability is increasing, which poses a notable risk for needed change and innovation [34]. A large majority of Western European project management practitioners has a higher education degree, between 73% and 85% [2].

Higher education for several professions like Information and Communication Technology (ICT), Engineering and Construction need to teach their students project management [6, 20, 30, 31, 35, 38]. Project management is offered as a significant component in a range of undergraduate and postgraduate academic qualifications including Construction, Engineering, and ICT, in line with the desire to make the higher education studies more relevant to the reality of the workplace. This workplace shows a steadily growing part of the work being undertaken in projects. More and

S. A. Nijhuis (✉)
Utrecht University of Applied Science, P.O. Box 182, 3500 AD Utrecht, The Netherlands
e-mail: steven.nijhuis@hu.nl

R. Cuevas et al. (eds.), *Research on Project, Programme and Portfolio Management*, Lecture Notes in Management and Industrial Engineering,
https://doi.org/10.1007/978-3-030-60139-3_18

more graduates will become part of projects and a growing number of them will be asked to lead them. Project manager is a realistic career path, which is comparable to a (line) management career path [9, 10, 18, 32].

Berggren and Söderlund [4] conclude that there is critique on how project management is taught, but a solution is not offered. Students need more competences than only knowledge of methods and tools [32]. But project management encompasses a very diverse set of competences. And the amount of room in a curriculum dedicated to project management is not without limits [10]. Resources are wasted by not teaching correctly. The necessary transition to a more sustainable world is compromised if higher education delivers graduates without the proper skills needed.

In order to get to a sustainable teaching of project management, choices have to be made, preferably based on research. Little research has been done on what actually to incorporate in project management education that aims at the inexperienced project manager. There are certification standards for project management, but they do not have a strong foundation in research [8], and at best a weak link with project success [21].

Although it is likely that different contexts (like ICT versus Construction) and different phases (initiating versus implementing) demand a different emphasis in learning project management, this is hardly reflected in the research on competences or learning environments. This article presents an overview of current insights on the design of project management curricula, highlighting insights from articles published in the last seven years on this subject, presenting previously unpublished comparisons and previously unpublished results. It will compare the offerings of commercial and higher education on project management learning. Next important and critical competences are discussed. Exploring context effects on critical competences shows unexpected results.

18.2 Offerings of Project Management Education

A total of 181 commercial courses, all offering a "standard" program for teaching project management in the Netherlands is analyzed [25]. Three types of purpose were found: general development (59%), learning a specific method like Prince2, Agile, or Scrum (22%), and preparing for general certification like PMP, IPMA D or IPMA C (19%).

The commercial offerings have a habit of not testing, especially the development courses do not mention testing at the end (73%). Among the commercial offerings are some higher education institutes that sell specialized courses or parts of their curriculum on a commercial basis (28% of all offerings).

The big majority (96%) of offerings targets junior or inexperienced project managers, is open for all contexts (86%) and is also "beneficial" for people working in projects, team leaders of people working/leading projects, and sometimes also for secretaries, accountants, consultants and/or sponsors. The primary teaching mode is face to face (78%). It is common to see marketing texts like "ability to run projects successfully" and "increasing project success".

Following this type of analysis, the offerings of project management education in higher education is discussed [26] using three sources: scientific literature on actual offerings (36 articles), the Dutch higher education (264 courses and programs), and the offerings of a group of reference institutes (33 courses and programs). All the reviewed offerings target the inexperienced project manager, sometimes with a line of courses. This is comparable with the commercial offerings. Project management in higher education is usually linked to the context of the study, with only a few elective courses open for all. The primary mode of teaching is students sitting in class listening (95% for the reference institutes, between 65% and 80% for the others). Higher education does test at the end, with a focus on assignments (68–81%) and (multiple-choice) sit down tests (43–63%). Scaling the promised level after the course on five options: unknown, knowledge of, insight in, experience with and ability to do project management reveals marketing like results. More than 40% of the courses claim experience in project management or even the ability to do project management (some even add "successfully in every context"). Only reference institutions are relatively modest, claiming roughly the same volume of knowledge & insight as experience & ability. The rest is not so modest in their claims. Marketing should not be a factor here, these course descriptions are for enrolled students. Marketing does play a role in program descriptions, used to get students into the curriculum. The promised level in program descriptions is indeed higher, albeit slightly.

On first sight higher education courses on project management seem to be akin to commercial courses for inexperienced or just started project managers. The descriptions are studied to determine which project groups and project phases get attention in the courses, assuming that when mentioned it is a focal point in the execution of the course. Another similarity can be noted based on this scrutiny: neither commercial nor higher education courses refer to a scientific research paper for the selection of competences, subjects or phases.

Of the project phases, the planning phase gets most attention, being mentioned by four out of five courses in all types (see Table 18.1 for more details), and being the only one appearing in more than two-thirds of the descriptions for all types. The second phase is controlling, mentioned approximately half the time in all types. Only

Table 18.1 Comparing offerings of project management on phases (previously unpublished)

Phase	Literature (%)	Reference (%)	University (%)	University of applied science (%)	Commercial (inexp. PM) (%)
Initiating	53	29	18	38	39
Planning	*89*	*95*	*77*	*87*	*83*
Implementing	42	48	36	38	29
Controlling	53	48	54	55	49
Closing	47	14	8	25	26

Table 18.2 Comparing offerings of project management on subjects (previously unpublished)

Subject	Literature (%)	Reference (%)	University (%)	University of applied science (%)	Commercial (inexp. PM) (%)
Communication	*53*	21	12	25	*51*
Cost	40	26	49	40	*64*
Integration	27	–	7	2	3
Procurement	20	21	22	5	4
Quality	*53*	16	27	30	*51*
Resource (incl. team)	*53*	32	29	30	*51*
Risk	*73*	*58*	*63*	*54*	*65*
Scope	*67*	21	34	39	35
Stakeholder	47	21	27	40	36
Time	*73*	42	*51*	44	*59*

in the lower scoring phases initiating, implementing and closing real differences seem to appear. Note that in literature on actual teaching of project management all phases are mentioned quite often compared to others.

Regarding the project management subjects (Table 18.2), the scoring is lower with only risk being mentioned by all types in at least half of the descriptions. The rest of the groups are less mentioned, with the exception of time, appearing in more than half of the descriptions of literature, universities, and commercial offerings. For the rest only literature and commercial seem to be just mentioning almost everything, with six out of ten subjects mentioned in more than half of the course descriptions.

Even at closer look, the commercial and higher education courses on project management seem to be not only targeting the same group, but also giving attention to the same subjects and phases. The biggest noted difference is that higher education actually tests more. If the commercials test, students usually either write an essay or do a knowledge test, which again is similar to the primary forms of testing in higher education.

Testing in higher education is usually done to award study credits, not to test whether students learned something. No references of measurement of learning gains have been found in the descriptions of commercial nor higher education project management courses. A renowned publication on measuring such effects states that in order to determine whether students learned something a pre- and post-test should be administered [16, 17]. The literature on actual teaching of project management does mention measurements of learning effects (20 out of 36), but only a small minority (4) actually involves pre- and post-tests. Almost all other measurements of effects are based on variants of Students Assessments of Learning Gains or SALG [37], or "asking the students". Self-assessment has been found inaccurate [39] and asking students whether they learned something will trigger a positive answer [33].

Further research into SALG using pre- and post-tests shows that those results are unreliable at best [24, 29].

Project-based learning (PBL) deserves a special note. Especially in literature, this method of learning is promoted as the instrument to learn project management, as is its cousin service learning, where students spend time on a project beneficial to society such as fundraising events or repairing damaged homes. Several authors note the dangers of project-based learning [3, 12, 41]. Project-based learning is not a guarantee for learning project management, on the contrary, as several examples in literature illustrate [13–15]. Moreover, a project in PBL is not a normal project since time and money are not factors that can be negotiated [22]. Next to this critique, it is noted that what is called a "project" in PBL is somewhat inflated, even calling the writing of an essay a project [26].

To summarize, project management education claims a relatively high skill level, but seems to be unable to live up to that claim. The focus is very much on the planning side of project management and on the risk management and time management, therefore not addressing the breadth of the profession. Only literature showed some efforts to review whether students or participants actually learned something, mostly using an unreliable method. The promoted method of PBL can be questioned on effectiveness.

18.3 What to Teach

In 2012, an early effort was undertaken to aggregate several studies in project management competences [22]. It showed that studies on project management competences did not share a common basis. The need for such a common basis before aggregating studies was underlined and a basis, suggested on the IPMA World Congress in 2014, was published in 2015 [30]. The question asked in studies using surveys among practitioners are variants of "what competences are important". Only a few suggest the notion of criticality [1, 19] but do not explore that notion. A first attempt at proving the importance of the use of criticality fails due to lack of sufficient comparable literature [27]. The difference between critical competences and important competences is illustrated in Table 18.3.

Analyzing 30 previous studies on competences using a taxonomy to categorize them reveals differences between 16 competences found important (see Table 18.4) [28]. Participants in nine focus groups of experienced Dutch project managers discussing criticality marked the critical competences on the competences of the

Table 18.3 Examples of important and critical project manager competences (previously unpublished)

Important competences	Critical competences
Basic computer skills	Political savvy
Technical (domain) knowledge	Organizational knowledge
Writing skills	Listening skills

Table 18.4 Competences found important in previous competence studies, distilled from [28]

	Competence	Important in (%)
1.	Leading	82
2.	Listening skills	80
3.	Oral communication	75
4.	Conducting meetings	75
5.	Problem-solving	73
6.	Written communication	71
7.	Trust worthiness	67
8.	Decisiveness	60
9.	Monitoring	59
10.	Initiative	57
11.	Decision-making	56
12.	Expectation management	56
13.	Team building	55
14.	Goal setting	50
15.	Decision delegation	50
16.	Customer focus	50

taxonomy, answering the question "in which competences do you consider to be better than your team members?" This produced 21 competences found critical (Table 18.5) [28]. The list of critical competences shows some overlap with important competences, but more differences.

The type of question asked has a big influence on the outcome, where criticality has a better potential of defining the essence of the profession than important. It is advised to use critical instead of important. Other recommendations based on the analysis of previous research is to make use of standardized lists of competences instead of the usual clean slate methodology and to provide definitions of competences in surveys [28].

18.4 The Influence of Context

Textbooks and global certification systems seem to suggest there is a universal competence profile. Nonetheless, a standard work on what should be part of the curriculum of project management suggest context should be taken into account when designing courses [40]. Several scholars find differences geographically [42], between types of project [5] and industry types [7].

Extra analysis on previous research shows that type of industry and geographic region do not change the surveyed and reported important competences [28]. Contrastingly, the agreement among the focus groups on what is critical seems to

Table 18.5 Competence marked critical by focus groups, distilled from [28]

	Competence	Critical by (%)
1.	Personal responsibility	71
2.	Organization knowledge	65
3.	Initiative	62
4.	Composure [under stress]	62
5.	Listening skills	59
6.	Motivating by persuasion	59
7.	Expectation management	58
8.	Resilience	58
9.	Position knowledge	58
10.	Cooperation	58
11.	Trust worthiness	57
12.	Analytical thinking	57
13.	Approachability	55
14.	People reading	54
15.	Tolerance of ideas	54
16.	Leading	52
17.	Decisiveness	52
18.	Team building	52
19.	Political astuteness	52
20.	Self-development	51
21.	Creative thinking	51

be low. The number 7 critical competence "expectation management" was marked in one group by only a quarter of the participants. Even bigger differences can be found, e.g. "decisiveness" (number 17) scores between 0% and 100%. These and some other examples are shown in Fig. 18.1.

These differences could be due to difference in context. Ruling out geographical differences (all focus groups are Dutch oriented), the focus turns to other differences like type of project and type of organization. Two focus groups share project type and type of organization (and even a common professionalization track for project managers). Their scoring and the scoring from the other focus groups show differences—as expected. But the differences in scoring between those two focus groups are much greater. Context is more complex than just geographical, project type and organization type.

Whereas the focus groups show little "convergence" on critical competences, they do show convergence on critical processes [27]. Although the focus groups show differences between them on critical competences, the average for junior project managers seems to be identical with the average of the participants in the focus groups, comprising of senior and experienced project managers [25], as illustrated in Fig. 18.2.

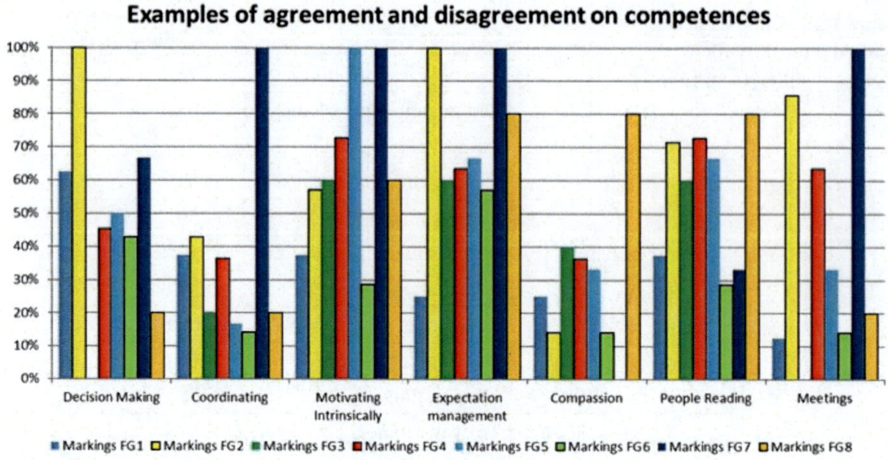

Fig. 18.1 Comparing criticality scores over focus groups (previously unpublished)

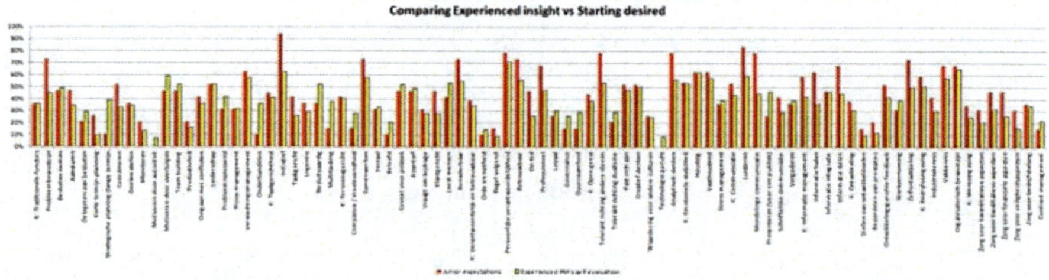

Fig. 18.2 Comparing averages experienced and junior project managers (from [25], by author, published open access at https://www.atiner.gr/papers/EDU2016-2154.pdf, free to share and adapt)

On critical processes, there is a hint of differences between junior and experienced managers: "Junior project managers" appear to be managing projects where stakeholder management is much less important. The focus on controlling the project appears to be more important for junior project managers' [25].

18.5 Response Group

The necessary insights of the response group are rarely addressed. Can students judge whether the setup of a course is indeed good for learning project management? One can doubt that, yet this is frequently done [26]. It is assumed that what comes into job advertisements is a good description of the actual task necessities [7], even without mentioning this assumption. When looking at samples in the list of previous research [28], a large variety of people with opinions on project manager competences can be observed participating in previous research. When "only" project managers are used, years of experience usually varies within this respondent group. When junior project managers have different critical competences or critical processes compared to experienced or senior project managers, mixing these groups in the results will not lead to insight for project management education.

Determining the right group for your research is an intricate question [23]. Especially when trying to determine the right learning goals for inexperienced project managers. There are differences in opinion even between students with internship experience and recently graduated students [11]. A junior project manager could lack the experience needed for a complete picture. An experienced project manager has that picture, but could lack insight into current demands on junior project managers. A sponsor or team leader is not actually performing the tasks. Equal objections can be made for consultants, team members and trainers of project managers. If they share the same opinion, that would make a strong case.

18.6 Conclusion

At this moment project management education is not based on research, teaching and assessment methods do not match the learning goals stated, even when project-based learning is promoted. There is very little proof of actual testing whether learning goals are achieved. There is no proof that the teaching of project management in this form is sustainable.

Previous research into competences cannot be used as a basis for teaching project management, since they promote general important competences over critical competences. Research into critical competences cannot be used as a basis, since it involved experienced project managers, reflecting on themselves. Notwithstanding the striking resemblance between critical competences for experienced and junior project managers, the high context dependability implies a high uncertainty of the correctness of any chosen critical competence for any specific context. We could be teaching them incorrect competences for their future employment. A process focus could be the solution for this dilemma. Further research is needed, giving attention to the correct respondent group, to assure sustainable teaching of project management.

Compliance with Ethical Standards The information in this chapter is a synthesis of previous research, complying with the ethical standards of thesis research of the Twente University in The Netherlands. Among those standards are strict anonymity, informing participants before data gathering, voluntary participation and the ability to opt out of participation at any time. Other information used in this research was publicly available to anyone (e.g. the information on project management courses). Scientific articles cited were obtained legally through library licenses.

References

1. Arras People, Thorpe J (2010) Project management benchmark report
2. Arras People, Thorpe J (2015) Project management benchmark report
3. Ashraf M (2004) A critical look at the use of group projects as a pedagogical tool. J Educ Bus 79(4):213–216
4. Berggren C, Söderlund J (2008) Rethinking project management education: social twists and knowledge co-production. Int J Proj Manage 26(3):286–296. https://doi.org/10.1016/j.ijproman.2008.01.004
5. Besner C, Hobbs B (2012) Contextualization of project management practice and best practice. Project Management Institute, Newtown Square, Pa
6. Car Z, Pripuzic K, Belani H (2010) Teaching project management to graduate students of electrical engineering and computing. Tech Technol Educ Manag TTEM 5(1):73–81
7. Chipulu M, JNeoh JG, Ojiako U et al (2013) A multidimensional analysis of project manager competences. IEEE Trans Eng Manage 60(3):506–517. https://doi.org/10.1109/tem.2012.2215330
8. Crawford L (2004) Global project management body of knowledge and standards. In: Morris PWG, Pinto JA (eds) The Wiley guide to managing projects. Wiley, Hoboken, NJ, pp 1150–1196
9. Crawford L, Morris P, Thomas J et al (2006) Practitioner development: from trained technicians to reflective practitioners. Int J Proj Manage 24(8):722–733. https://doi.org/10.1016/j.ijproman.2006.09.010
10. Ellis R, Thorpe T, Wood G (2003) E-learning for project management. Proc Inst Civ Eng Civ Eng 156(3):137–141
11. Geelen B (2019) Projectmanagement in het onderwijs; onderscheidende competenties voor opleidingen Sportkunde. Master, Utrecht University of Applied Science
12. Graaf Ed, Kolmos A (2003) Characteristics of problem-based learning. Int J Eng Educ 19(5):657
13. Heineke J, Meile L, Liu LB et al (2010) Teaching brief: project flip: a project management case/exercise experience. Decis Sci J Innovative Educ 8(1):113–127
14. Jewels T, Ford M (2004) A single case study approach to teaching: effects on learning and understanding. J Issues Inf Sci Inf Technol 1:359–372
15. Jollands M, Jolly L, Molyneaux T (2012) Project-based learning as a contributing factor to graduates' work readiness. Eur J Eng Educ 37(2):143–154
16. Kirkpatrick D (1959) Four-level training evaluation model. US Train Dev J
17. Kirkpatrick D (1996) Great ideas revisited. Techniques for evaluating training programs. Revisiting Kirkpatrick's four-level model. Train Dev 50(1):54–59
18. Martin A (2000) A simulation engine for custom project management education. Int J Proj Manage 18(3):201–213. https://doi.org/10.1016/s0263-7863(99)00014-9
19. McHenry RL (2008) Understanding the project manager competencies in a diversified project management community using a project management competency value grid. Ph.D., Capella University
20. Mengel T (2008) Outcome-based project management education for emerging leaders—a case study of teaching and learning project management. Int J Proj Manage 26(3):275–285. https://doi.org/10.1016/j.ijproman.2007.12.004
21. Morris PWG, Crawford L, Hodgson D et al (2006) Exploring the role of formal bodies of knowledge in defining a profession—the case of project management. Int J Proj Manage 24(8):710–721. https://doi.org/10.1016/j.ijproman.2006.09.012
22. Nijhuis SA (2012) Learning for project management in a higher education curriculum. In: Project management institute research and education conference 2012, Limerick, Ireland
23. Nijhuis SA (2015a) Critical engagement of previous research. In: Pasian B (ed) Designs, methods and practices for research of project management. Gower Publishing, Burlington, USA
24. Nijhuis SA (2015b) Project management tuition or training, can we assess the added value of them? Proc Soc Behav Sci 194:146–154. http://dx.doi.org/10.1016/j.sbspro.2015.06.128

25. Nijhuis SA (2016) The gap between what is needed and offered in project management education. In: ATINER'S conference paper series, No: EDU2016-2154, Athens, Greece, vol EDU2016-2154
26. Nijhuis SA (2017) Exploring project management education. Eur J Soc Sci Educ Res 1:44
27. Nijhuis SA, Kessels JWM, Vrijhoef R (2015) The importance of criticality in (project management) competence research. In: 19th international conference on engineering education: new technologies and innovation in education for global business, Zagreb, Croatia, 20–24 July 2015, pp 690–697
28. Nijhuis SA, Vrijhoef R, Kessels JWM (2018) Tackling project management competence research. Proj Manage J 49(3):62–81. https://doi.org/10.1177/8756972818770591
29. Nijhuis SA, vrijhoef r, kessels jwm (2015a) measuring learning gains in project management. In: 19th international conference on engineering education: new technologies and innovation in education for global business, Zagreb, Croatia, 20–24 July 2015, pp 690–697
30. Nijhuis SA, Vrijhoef R, Kessels JWM (2015b) Towards a taxonomy for project management competences. Proc Soc Behav Sci 194:181–191. http://dx.doi.org/10.1016/j.sbspro.2015.06.132
31. Nooriafsha M, Todhunter B (2004) Designing a web enhanced multimedia learning environment (WEMLE) for project management. J Interact Learn Res 15(1):33–41
32. Pant I, Baroudi B (2008) Project management education: the human skills imperative. Int J Proj Manage 26(2):124–128. https://doi.org/10.1016/j.ijproman.2007.05.010
33. Pligt J, Blankers M (2013) Survey-onderzoek. De meting van attitudes en gedrag. Den Haag Boom Lemma uitgevers 2013
34. PMI (2017) Project management job growth and talent gap 2017–2027
35. Reif HL, Mitri M (2005) How university professors teach project management for information systems. Commun ACM 48(8):134–136
36. Sexton P, Foley E, Wagner R (2019) The future of project management: global outlook 2019
37. Seymour E, Wiese D, Hunter A et al (2000) Creating a better mousetrap: on-line student assessment of their learning gains. In: National Meeting of the American Chemical Society, San Francisco, CA, 27-03-2000
38. Stoyan R (2008) "PM for all™"—intensive small group teaching in leadership and PM, for many students at low cost. Int J Proj Manage 26(3):297–303. https://doi.org/10.1016/j.ijproman.2008.02.006
39. Symons AB, Swanson A, McGuigan D et al (2009) A tool for self-assessment of communication skills and professionalism in residents. BMC Med Educ 9(1)
40. Task Force on PM Curricula (2015) Project management curriculum and resources. Project Management Institute, Inc., Newtown Square, Pennsylvania, 19073-3299, USA
41. Volkema RJ (2010) Designing effective projects: decision options for maximizing learning and project success. J Manage Educ 34(4):527–550
42. Zwikael O, Shimizu K, Globerson S (2005) Cultural differences in project management capabilities: a field study. Int J Proj Manage 23(6):454–462

Chapter 19
Do Educators Need to Develop Project Management Competencies for a Rapid and Sustainable Development of New Academic Development Programs?

Adán López Miranda

Abstract Crashing projects to target ambitious dates is commonplace in highly competitive industries. To face these challenges, teams are trained in project management methods and competencies. In education environments like schools or universities these demanding conditions do not appear to be the case. Educators normally develop academic projects in small teams where the pace is comparatively slower. This paper presents a real case where a group of educators worked in a team to develop an on-line course in half the usual time. The team was composed of educators (including the author), web designers, programmers and marketing people and applied a project-accelerating method to face the challenge. Through participant observation, the author narrates the experience and suggests that educators' skills might be nurtured with project management competencies for a rapid and sustainable development of new academic programs.

Keywords Project management competencies · Educator competencies · Participant observation · Concurrent project management

19.1 Introduction

The Tecnológico de Monterrey is ranked 158 among the best universities in the world (QS Ranking, 2020) and began the production of on-line education programs since 1989. The university still plays a key role in this growing industry where there is fierce worldwide competition and companies such as Coursera, Udemy and Udacity appear to be the market leaders.

Designing, developing and launching a new on-line course can be a daunting endeavour. Some courses cover several learning modules, themes and subthemes requiring the student to spend more than 100 h reading lectures, answering questions and taking tests. These programs require teams of 10–20 persons who must develop

A. López Miranda (✉)
Tecnológico de Monterrey, School of Engineering, Av. Eugenio Garza Sada 2501, 64849 Monterrey, N.L., Mexico
e-mail: adlopez@tec.mx

R. Cuevas et al. (eds.), *Research on Project, Programme and Portfolio Management*, Lecture Notes in Management and Industrial Engineering,
https://doi.org/10.1007/978-3-030-60139-3_19

the project in time, at a budgeted cost and within the quality specified by the course material and learning outcomes. Typically, the teams are composed of educators, web designers, programmers, marketing people and the project coordinator. In all, the development of these programs consumes more than 10,000 h of effort. The more stringent the constraints, the higher the need for the correct application of project management competencies, tools and techniques.

Educators (professors or lecturers in the university) are key team members in these projects but have not been trained in project management and many of them do not consider it a necessary skill when creating learning content. There is a natural inclination towards intellectual work that depends on work done in a quiet, reflexive, research-intensive and autonomous environment. However, these less stringent conditions are changing because of the need of responding faster to the exigencies of a digitalized, global and fast-changing society.

A case is presented that involved a project to design, develop and launch an online project management (PM) program. The authorization to start the project was delayed for four months, but the target date to open it did not change. Therefore, the project had to be developed in half the normal time for this kind of program.

To face this apparently insurmountable challenge, PM techniques were applied using an accelerated method to develop activities. Doubts were raised as to the response and attitude that educators (in the role of authors and instructional designers) would take to this approach driven by efficiency. The story narrated here poses the following question: Do educators need to develop PM competencies?

19.2 Objectives

The main goal of this study is an ex post facto analysis of the competencies of a group of educators that were part of a team that developed a project in half the usual time. To achieve this ambitious objective, the team applied an accelerated method for crashing the schedule. The author of this paper took part in the project and applied the participant observation method to gather the data and clues about the educators' competencies in developing projects.

This paper is structured in the following way. First, the method applied to accelerating or crashing projects is explained. Second, a brief review of the literature is shown regarding PM and educator competencies as well as a brief description of the participant observation research method. Then, the case is explained followed by an ex post facto analysis of the educators' experience in the project. Finally, the research question, "Do educators need PM competencies?" is discussed and recommendations for sustainable education are given.

19.3 Theoretical Background

19.3.1 A Model to Accelerate the Speed of Projects

The method described here was part of the empirical results of a doctoral thesis aimed at explaining the relationship between PM, Concurrent Engineering (CE) and New Product Development (NPD) [11]. For this research, four case studies were developed to describe the similarities and differences between these academic field of studies. The main research questions were posited as: (a) Do CE, NPD and PM practices emphasize essentially the same aspects (practice and purpose) or are they distinctly different? (b) Are CE, NPD and PM competing or complementary practices? (c) Is one component or a precursor of the other? To answer these questions, an exploratory study was launched based on the multiple-case study methodology [6].

The data from the settings (companies) were obtained from three different sources: interviews, documents and observations. The people interviewed worked in activities related to project management, like project managers, heads of program managers and also heads of PM training programs. In sum, 36 persons were interviewed for an average of 2 h each.

Although complete and illustrative, the IDA Report R-338 [10] definition of CE is difficult to understand and operationalise because of its many concepts. Hence, based on the constant comparison method [13], data were analysed and a simpler definition of CE was proposed: "CE is an approach for designing products or services through the application of multidisciplinary teams working in parallel and converging on data and knowledge". [11].

This new definition is composed of three components that are shown in Fig. 19.1 in the form of a mind-mapping tree (hereinafter called the concurrent model).

Parallel development. It refers to the simultaneous problem-solving approach described by Clark and Fujimoto [3] among others. Downstream activities start before upstream activities finish, and this occurs amid an intense information exchange intended to avoid reworking as illustrated in Fig. 19.2.

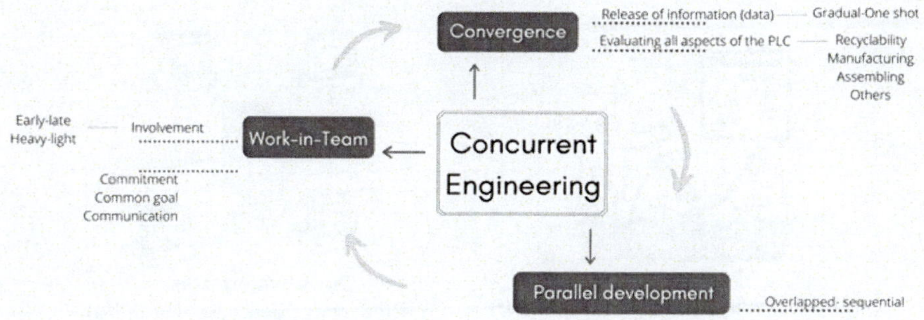

Fig. 19.1 Conceptual structure of concurrency [11]

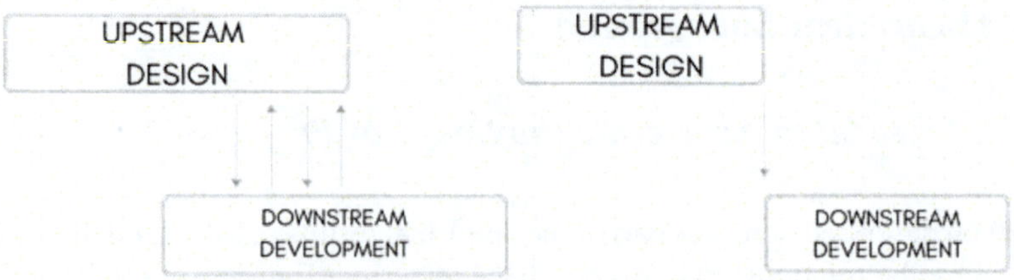

Fig. 19.2 Simultaneous problem-solving approach (left) and serial approach (right)

Work-in-team. It implies that the members of multidisciplinary groups work together to pursue a common goal. The relevant aspects that characterize effective teamwork include multidisciplinary problem-solving, common goals, commitment, collaboration and good levels of communication.

The project manager might compel (or inhibit) project teams, "against their natural inclination", to address simultaneous problem-solving, to join together early in the project or to enhance communication.

Convergence. This means that data and knowledge are concentrated in one physical or virtual space that is shared by the different team members. As shown in Fig. 19.3, four forms of convergence were observed during the case studies: meetings, co-location, on-line information and public boards. First, the teams converged or met to work together either as co-located teams (in Companies 1, 2 and 3) or in meeting rooms (in Companies 1 and 4). Companies 1, 2 and 3 used different tools to converge, such as instructions appearing on intranets. Several programs and project

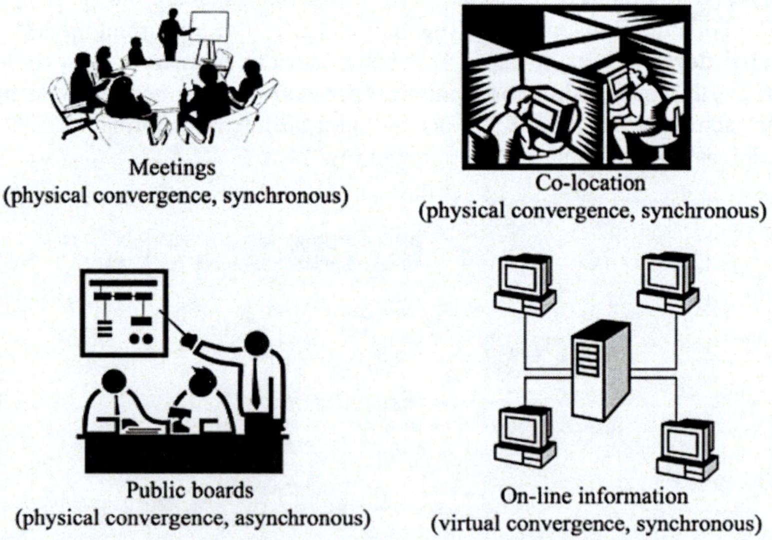

Fig. 19.3 Four types of convergence on information [11]

managers used the typical "war rooms" with public boards hanging on walls that contained project information to be shared with all team members and project stakeholders. Gantt charts, design sketches, product breakdown structures and other items were displayed. These boards were visual instruments that converged information as well. Regardless of the form of convergence, the goal was the same: to share information in order to expedite decision-making.

19.3.2 Project Management Competencies

The concept of competencies has been around for a long time. First, it referred to labour and then to general professional qualifications. Four characteristics of competencies are.

Labour-oriented. Generically, a competency is defined as a set of skills, abilities, knowledge and attitudes that are used to solve problems assertively or to provide a sound response to a new or specific situation [4].

Professional career-oriented. Professional competencies are a set of knowledge, skills and aptitudes that are needed when exercising a profession, solving problems of the same profession autonomously and flexibly, as well as having the ability to assist in the professional context and in the organization of work [2].

Project management-oriented. The application of knowledge, skills and abilities to achieve the expected results [9].

The IPMA ICB 4.0 [9] divides the PM competencies into three areas: Perspective through which the team members interact with stakeholders outside the project; People, the human inter-relationships inside the project; and Practice, the technical skills to deliver results.

According to [9] (p. 28), the competency areas are broken down into 29 elements: perspective competencies (5 elements), people competencies (10 elements) and practice competencies (14 elements). Examples of elements of competencies classified in these areas are.

People. Personal communication, leadership and teamwork

Practice. Design, time, scope, quality and finance

Perspective. Strategy, governance, power and interest.

For a complete list of elements of competency, the reader is invited to review the IMPA ICB 4.0 standard [9].

19.3.3 Educators' Competencies

Wesselink and Arjen [15] studied the competency profile for educators in organizations. The results of this research challenged educators to improve the following project management-related competencies.

- *Goals, objectives and benefits*: The education practitioner must focus the project results in satisfying customer needs.
- *Results orientation*: Despite creativity, educators have to deal with limited time and hand over deliverables.
- *Resourcefulness*: The education practitioner has to be able to solve problems using creative solutions under limited budget, resources or even time.
- *Relationships and engagement*: The educator should build professional networks in and outside education circles to keep abreast of the advance in his or her subject matters.

Ibrahim et al. [8] included lecturers' competencies in their study and related them to student satisfaction: knowledge of subject, clarity of presentation, punctuality, interaction with students, teaching creativity, course objectives, learning outcomes, assignments, class presentations, examinations, class activity, lecturer notes, student evaluations and class preparation. Clearly none of these competencies are related to PM. Such competencies would include course planning and progress tracking, for instance.

The University of Missouri website lists core competencies for its faculty and staff [5], some of them are naturally linked to academic skills like *education delivery*, *educational programming*, and *knowledge of subject matter.* Some others are related to PM competencies, like *communication, organizational management* and *leadership.*

This brief overview of lecturers' competencies reveals that pedagogical and organizational skills are required for university staff and faculty to complete their professional profiles. This review of the literature was exploratory, not exhaustive, because the goal of this paper is to answer a question based on the lessons learned during a real case.

19.3.4 Participant Observation

Marshal and Rossman [12] defined participant observation as "the systematic description of events, behaviours and artefacts in the social setting chosen for study".

According to Schensul et al. [14], two reasons for applying participant observation are: (a) to aide an investigator to obtain the feel for how things are organized, how people interrelate and the cultural aspects; (b) to show what the cultural members consider important in a socio-political environment.

Bernard [1] considered that the method makes it possible to gather different types of data; it facilitates involvement in sensitive activities to which the researcher would not normally be involved. Additionally, participant observation gives the researcher a better understanding of what is happening in the field of study.

Gold [7] described four observation stances that categorize the level of involvement and the extent to which the members acknowledge the role of the researcher: (a) Complete participant, total involvement and the role as researcher is hidden to the group; (b) Participant as observer, member of the group and the members are aware of the researcher role; (c) Observer as participant, the researcher may participate in some activities but his main role is as researcher and this is known by group members; (d) Complete observer, null interaction with the group and the role is unknown to participants.

The setting for this study was a project in which the subjects of the study were educators interacting in a competitive environment. The writer of this article participated in the project as leader author, not initially as researcher. In this sense the stance corresponds to complete participant. After the project was finished, it was decided to collect and analyse data in an effort to learn from the experience and in this sense the stance changed to "participant as observer" developing ex post facto research.

19.4 The Project

The need to design a new on-line PM seminar or *Diplomado*[1] arose in 2017 because the actual had been in the market for about six years. Its look and feel and user-interactive means were out of date. Additionally, there was the need to align the contents to a new version of the PM standard.

The new program would be competence oriented with a new user interface and would be composed of seven learning modules called journeys. Each module would require three hours of user involvement to acquire the learning outcomes. Translating this unit into an effective project work corresponds to working on a 30-page, full-text Word document. Traditionally, developing each trajectory starts with writing the original text, which is done by academic authors or educators. Then the text is reviewed by instructional designers (academic profile) and passed to graphical designers. When the trajectory content and story board are completed, programmers convert images and texts into interactive on-line media resources so the user can participate in the learning experience. Finally, the coded program is uploaded by IT experts onto the server where the program will be accessed. In summary, an entire group of specialists from different disciplines were involved in the Diplomado design and development. Additionally, other experts joined the project to enhance the attractiveness, selling and operation of the program. In all, approximately 20 persons worked directly on the program, led by a program coordinator who played the role of project manager.

[1] Training program of at least 96 h where a participant obtains a Diploma if successfully finish it.

The launch date for the first trajectory was the first week of April 2018 and work on the project was expected to start in early September 2017. However, budget restrictions and a complex decision-taking process delayed the authorization to start the project until the first week of December 2017—a three months' delay and that did not include permission to move the program launch date. The endeavour appeared to be nearly impossible according to experts with long experience in developing on-line programs.

Due to the stringent conditions, the author of this paper proposed applying the concurrent model of PM. The three main approaches of parallel development, a platform for converging on information, and a better way to work as a team would be implemented through this accelerated method (see Fig. 19.1).

Parallel development. The design of the journeys would be executed with five authors and four instructional designers. The authors would have to work during December to pass the learning content to instructional designers in January. It became clear that authors would not finish their journeys in one month because of the long content and the holydays. It was then decided to modularize the journeys in shorter courses (3–4 courses for each trajectory). Figure 19.4 presents a schematic chronogram used to execute activities in parallel and assign resources. Note that:

- The essence of parallel development is shown in Course 1 activities where downstream actors started before upstream actors finished.
- The quality review activities took less time in general because of the intense communication among team members.
- The marketing and sales activities started almost at the beginning, unlike the normal practice of started near the end of the programme, when the whole content was ready.

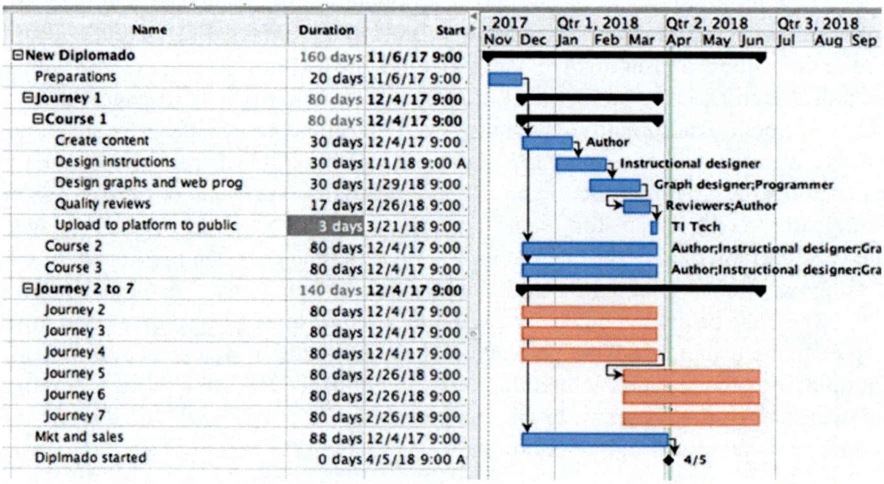

Fig. 19.4 Parallel approach executed by courses instead of by journeys

Unlike the serial processes followed in the past, this time the authors (upstream) shared content information with industrial designers (downstream) throughout the entire design phase. Instructional designers made preliminary reviews and gave feedback to the authors. This information exchange in both directions improved the quality of the content and reduced the "pass the ball" effect.

Efficiency in parallel development requires intense communication and a person's willingness to work with preliminary information that is still being developed [3]. At the beginning, the upstream actors of the development were not familiar with this new condition. As a result, receiving constant feedback from downstream actors irritated them in some ways. Some instructional designers still requested that the learning content be finished before they started their part of the job. However, they soon realised the convenience of working in this new way instead of passing and receiving batches of information.

Convergence. Traditionally, authors work and store their information independently and pass it through e-mail attachments to instructional designers who in turn work on the information personally and so on. This is the batch process. Although this method is convenient for the individuals, it can result in confusion, missing information, lack of integration and poor documentation. Hence, in this case, information was stored and shared in a central repository. This decision was productive, but it was nonetheless difficult to implement.

The first barrier was technological. The shared content file was located on a Google Drive platform and the team members were requested to install a local Google Drive on their computers to synchronize the files instantly. Several computers and laptops were not synchronized due to incompatible operating systems, access permissions or unknown causes.

The second problem was resistance to share. Some participants were resistant to being edited or afraid of having important information deleted. Others considered their content not ready to be shared and read by others.

The third problem was lack of expertise. Many persons in this group had not worked with shared platforms, so they did not even know that files were automatically synced in their local computers and in the central file. Therefore, they created backup files.

These initial limitations meant that the actors preferred to work individually; after they accumulated a batch of information, they uploaded it. This caused delays because other persons could not start working without the information. Fortunately, the group leaders maintained their constancy of purpose. Also, the higher rate of productivity by those who shared information produced a positive effect and drove the change. There were also rules in place about sharing information safely. For instance, the information on the central file was backed up weekly.

Instant group communication was introduced using mobile apps. Through these applications, the teams received instant messages regarding questions, criteria applied or how to solve problems. At first, the frequent text messages disconcerted people who were not accustomed to it, but they soon realised the benefits.

Instructional designers and web programmers were relocated to sit together to ensure quick information exchange. As proposed by the model, co-location is a means of converging physically on information (see Fig. 19.3).

Work-in-Team: When developing on-line programs, the producing teams traditionally treated content experts (professors and lecturers) as providers because they were professors or lecturers whose job was to deliver learning content. The role of delivering content was changed in this project to allow content providers to become part of the development team. Hence, they took part in commercial decisions like naming the program, marketing information and even pricing. They could also raise comments regarding the templates proposed by instructional designers.

Regular meetings were held involving all actors in the stages of development to avoid the "broken telephone" syndrome. These meetings improved clarification, mutual understanding and trust. All team members were invited to team-building sessions and to celebrations of achieving important milestones.

The early involvement of the marketing and sales department was noteworthy. They started by surveying students of the previous *Diplomado* in order to get insights on how to improve the new one. Meetings were held in which the entire team, including marketing experts, decided on the price, the name of the program and suggested the content outline. Almost from the beginning of the design stage, advertising was started. Also, a webinar was cast showing previews with already designed graphic materials.

19.5 Results

The on-line program opened to the public on the day planned for journey 1, while the others were developed in parallel and also opened on time. Registration reached record numbers. There were 470 participants, more than twice the expected goal of 200. The development cost was on budget. The quality of learning content was improved because a much more attractive and interactive user interface was in place. The satisfaction index, measured through the customer survey revealed an average number of 8.5 out of 10, similar to the old *Diplomado.*

The accelerated model approach to this project proved to be effective and allowed the team to achieve the time-to-market target in record time. The quality of the learning content was improved due to an intense production cycle: reviewing texts in parallel, at the same time, not in batches. Documentation was improved due to the fact that all files were in the same repository, not stored in different versions or under different document names. Working as teams was clearly an improvement over the normal practices, although more could have been done to reinforce team spirit. Team-building sessions or an initial kick-off event intended to motivate the whole team to achieve an aggressive target could have been included.

Two actions were applied to collect the lessons learned: an internal survey to the members of all participating organizational units and a feedback session with the main team members. As shown in Table 19.1, the opinions of the educators—content developers and instructional designers—were gathered and classified within the three categories of the concurrent model.

As can be seen, positive comments and things to improve about the same issue were highlighted by the academic members. For instance, many opinions were in favour of using Google Drive to instantly share documents, while others felt that there should have been more openness to share information. Regarding parallel development, one opinion was that the process was agile, while another said that there should have been a better understanding of the concurrent model. In the end, some educators accepted the new way of working; others did not. It was difficult to sustain a controlled flow

Table 19.1 Educators' responses to use of the concurrent model

Positive aspects	Need to improve
Comments about parallel development (PM competence: time)	
Agile process	A better understanding of the concurrent model
Good planning considering the short time we had	Experts must deliver the content complete before starting the instructional design
Excellent collaboration between content experts and instructional designers to conceptualize the program design	A lack of validation of the course elements because sometimes we dealt with too many elements at the same time
	It was difficult to work with different modules at the same time
Comments about concurrence or convergence	
The creation of the Whatsapp group for instant real-time communication	We should be more open to share documents while we are developing it
Agile communication	
Using Google Drive to have updated documents all the time	
Excellent communication flow	
The use of Google Drive together with a usage protocol	
Direct contact with content experts helped to rapidly clarify doubts	
Working collaboratively with Google Drive	
Comments about work-in-team	
Real commitment of all team members towards the accomplishment of the goal	Better communication between content developers and instructional designers
Working together with the content experts	Communication was not homogeneous among all team members
My comments about the content were immediately considered	The different team members' roles should be more clearly communicated
We all contributed to the achievement of the goal	Improving communication

and level of communication in an environment where a rapid pace for completion of a project was a high priority.

19.6 Discussion and Recommendations

This article presents a case study observed through the point of view of a participant. It was a project in which team members were required to work in parallel, converging on information and working with a clear orientation to work as a team. These conditions appear to be commonplace for professionals working in highly competitive industries and trained in PM. However, they are not usual for educators working at a university. One issue that emerged was the question, Do educators need to develop PM competencies?

A brief review of the literature shows that most of the PM competencies listed in the [9] standard are not part of the lecturers' core or complementary skills. Answering whether PM skills should be standard for educators' competencies requires more research, but the evidence from the case presented here indicates that educators participating in a challenging project needed these abilities. Time, teamwork and effective exchange of information as well as the creation and storage of documentation, were essential. This, of course, does not mean that professors might not have those abilities, but the fact is that these skills did not appear in the group. People working on a team project should at least be aware of PM competences.

Several educators interviewed during the feedback sessions recognized that to some extent PM competences were a key component of the success of the project. Others felt that the exigencies of working concurrently were not a safe, easy nor efficient method that was above all deficient in communication.

More research should be done in terms of including PM skills as competencies. It is clear that universities are getting involved in more competitive environments because open courses, on-line free programs and massive open on-line courses (MOOCs) put pressure on academics to develop faster and cheaper education initiatives that guarantee quality. Training, awareness and application of PM competencies are then recommended for educators in preparation for facing these new and demanding exigencies in a sustainable way.

Compliance with Ethical Standards All procedures performed in this study involving human participants were in accordance with the ethical standards of the institutional research committee of the School of Engineering at Tecnológico de Monterrey (Permission ALM 24-04-2020). Active participants of the project were informed in writing about the content of this article and their consent was obtained to publish it. Anonymity is kept since there is no direct or indirect information that might lead to identify any particular person or group thereby individual rights are not infringed.

References

1. Bernard HR (1994) Research methods in anthropology: qualitative and quantitative approaches, 2nd edn. Alta Mira Press, Walnut Creek, CA
2. Bunk G (1994) La transmisión de las competencias en la formación y perfeccionamiento profesionales en la RFA. CEDEFOP
3. Clark KB, Fujimoto T (1991) Product development performance, strategy, organization, and management in the word auto industry. Harvard Business School
4. CONOCER (2018) Consejo Nacional de Normalización y Certificación de Competencias Laborales. Available via http://conocer.gob.mx/preguntas-frecuentes/. Accessed 15 Aug 2019
5. Core Competencies for Faculty and Staff (n.d.) Extension University of Missouri. http://extension.missouri.edu/staff/core-competencies.aspx#gsc.tab=0. Accessed 8 June 2019
6. Eisenhardt KM (1989) Building theories from case study research. Acad Manage Rev 14(4)
7. Gold RL (1958) Roles in sociological field observations. Soc Forces 36:217–223
8. Ibrahim AZ, Owee TK, Sang CL (2014) Analysis on the relationship between lecturers' competencies and students' satisfaction. Int Educ Stud 7(1)
9. IPMA (2015) ICB 4.0 IPMA individual competence baseline. Version 4.0
10. Winner RI, Penell JP, Bertrand HE et al (1988) IDA: Report R-338 The role of concurrent engineering in weapons system acquisition
11. López Miranda A (2007) The application of NPD, CE and PM techniques to develop new products. Dissertation, University College London, London, UK
12. Marshall C, Rossman GB (1989) Designing qualitative research. Sage, Newbury Park, CA, p 79
13. Suddaby R (2006) What grounded theory is not. Acad Manage J 49(4):633–642
14. Schensul SL, Schensul JJ, LeCompte MD (1999) Essential ethnographic methods: observations, interviews, and questionnaires. Book 2 in ethnographer's toolkit. Alta Mira Press, Walnut Creek, CA
15. Wesselnik R, Arjen EJ (2011) Developing competence profiles for educators in environmental education organisations in the Netherlands. Environ Educ Res 13504622, 17(1)

Chapter 20
Sustainability in the IPMA 4LC Certification Process in Spain (2014–2019)

Ángela Paneque de la Torre, Vanesa G. Lo Iacono Ferreira, María José Bastante-Ceca, and Salvador F. Capuz-Rizo

Abstract The inclusion of concepts and sustainability principles in the standards of project management is part of the materialization of the integration of sustainability in this discipline. The objective of this research is analyzing if any of the factors evaluated in order to certificate a project manager, according to the International Project Management Association (IPMA), contain sustainability or sustainable development aspects. Moreover, an assessment of how these factors are considered and their relationship with the rest of the factors taken into account is developed.

Keywords Project management · Competences · Professional certification · Sustainability

20.1 Introduction

The integration of sustainability in project management is the ongoing trends in research within this discipline. A clear indicator of the influence of sustainability on project management is the integration into standards [13]. Therefore the inclusion of the concepts and sustainability principles in the standards of project management is part of the materialization of this integration.

Á. Paneque de la Torre (✉) · V. G. Lo Iacono Ferreira · M. J. Bastante-Ceca · S. F. Capuz-Rizo
Universitat Politècnica de València, Camino de Vera s/n, 46022 Valencia, Spain
e-mail: anpade@doctor.upv.es

V. G. Lo Iacono Ferreira
e-mail: valoia@upv.es

M. J. Bastante-Ceca
e-mail: mabasce1@dpi.upv.es

S. F. Capuz-Rizo
e-mail: scapuz@dpi.upv.es

V. G. Lo Iacono Ferreira
Universitat Politècnica de València, Plaza Ferrándiz y Carbonell 1, 03801 Alcoy, Alicante, Spain

R. Cuevas et al. (eds.), *Research on Project, Programme and Portfolio Management*, Lecture Notes in Management and Industrial Engineering,
https://doi.org/10.1007/978-3-030-60139-3_20

Previous literature reviews show a growing trend in the number of academic publications on the topic of sustainability and project management [1, 9]. However, according to Silvius [13], the attention given by the academic community to the relationship between sustainability and project management differs from how project management standards address this relationship.

The International Project Management Association (IPMA) offers an international, four-level certification scheme. The certification is based on the global competence standard, ICB. IPMA ICB standard describes the competencies that project, program and portfolio managers should have. When publishing this chapter, IPMA was upgrading from ICB version 3 [6] to ICB version 4 [7]. All certifications will be based upon ICB version 4 by the end of 2019, worldwide [8].

In Spain, the Certification Body of the Project Management (OCDP) is the specific entity authorized to operate this scheme and certification. It is validated by the IPMA-CVMB (Certification and Validation Management Board) and also accredited by the Spanish National Accreditation Body (ENAC) for Certification of Persons, according to ISO 17.024, in the category Project Managers [2]. The OCDP panel of assessors is made up of 16 project management professionals (of which 12.5% are women) certified as IPMA assessors.

Since level D does not require report and interview and only knowledge about competences is assessed and not competences involved in real projects, this study is limited to the certifications of levels A, B and C carried out by the Spanish Certification Body from 2014 to 2019.

20.2 Background

This study formulates the research question: *Have aspects related to sustainability been considered in the professional certification of project managers according to the IPMA standard ICB version 3? To what extent are they considered?*

The objective of this work is analyzing if any of the factors assessed in order to certificate a project manager, according to IPMA contain sustainability or sustainable development aspects. Moreover, an evaluation of how these factors are considered and concerning the rest of the factors taken into account is developed.

Økland [10] developed a study about the deficiencies between the measures suggested in the literature for the incorporation of sustainability in project management and what is carried out in practice. Regarding project management standards, specifically in the PMBoK Guide, the study shows that sustainability is mentioned as a contextual and external factor that the project manager should be aware of, because of the influence that sustainability policies and regulations may have on the project. According to Økland, the review of Eskerod and Huemann [4] on three project management standards such as the Guide to the Project Management Body of Knowledge (PMBOK 3rd edition), the International Competence Baseline (ICB 3.0) [6] and PRINCE2 4th edition, concludes that neither of them explicitly considers sustainability issues.

Eskerod and Huemann [4] emphasize that the standards selected are used all over the world as the basis for certification of project managers. Therefore, it can be construed that a significant amount of time is spent on learning these concepts by many people. These standards are some of the strong isomorphic forces existing in the project management profession [3]. Hence sustainability should be present in these standards.

Although the word sustainability was not explicitly mentioned in the ICB version 3, several authors have found references and implicit consideration of sustainability and sustainable development principles in this competency-based standard, as shown in Table 20.1.

According to Silvius et al. [15], the explicit recognition of the categories' behavioural and contextual competences fits the expected impact of sustainability concepts. They found some references to environmental and social aspects among the competence elements (CE) of the ICB3. For these authors, even the elements "Permanent organization" and "Business" include references to sustainable aspects in case the organization is adopting sustainability in its business strategy.

Eskerod and Huemann [4] analyzed three standards in project management comparing stakeholder management. They found that the ICB states to "include

Table 20.1 Sustainability references in the standard IPMA ICB version 3

	Silvius et al. [15]	Eskerod and Huemann [4]	Gareis et al. [5]	Silvius [14]
Technical competences				
1.02 Interested parties		x		
1.04 Risk and opportunity			x	
Behavioural competences				
2.06 Openness			x	
2.09 Efficiency			x	
2.12 Conflict and crisis			x	
2.14 Values appreciation	x		x	
2.15 Ethics	x		x	
Contextual competences				
3.05 Permanent organization	x			
3.06 Business	x			
3.08 Personnel management	x		x	
3.09 Health, security, safety and environment	x		x	x

interested parties' interests and expectations in the requirements, objectives, scope, deliverables schedule and costs of the project plan". They conclude that stakeholders are considered seriously. They also understood that stakeholders' expectations and needs are encompassed in the project interests.

The analysis of project management standards carried out by Gareis et al. [5], concludes that the strength of the ICB3 standard lies in the care for the project personnel (included in CE 3.08) and the consideration of values such as ethics (CE 2.15), openness (CE 2.06), social sensitivity (CE 2.14), fairness (CE 2.12), integrity (CE 2.15 and 2.12), transparency (CE 2.15 and 2.12), respect (CE 2.15 and 2.14) and efficiency (CE 2.09). They add project risk management (CE 1.04), which is a significant concern of sustainable development and is broadly referred to in the project management standards, being an additional implicit consideration of sustainable development principles.

In a Q-methodological study, Silvius et al. [14] are based on the ICB3 competence element 3.09, to include "Health and safety issues are checked" among the statements of the sustainability category for their Q-sort.

Paneque-de la Torre et al. [12] analyzed the principles and aspects of sustainability introduced in the IPMA ICB, and its change from ICB3 (2006) to ICB4.0, published in 2015. According to these authors, while in version 3, the concept of sustainability did not appear explicitly (although related concepts were discussed), in the IPMA ICB4.0, sustainability appears explicitly and becomes more prominent. Analyzing the frequency of occurrence of the word "Sustainability" in the different areas of competence (Practice, People and Perspective) of the IPMA ICB4.0 for the domain of project management, it is observed that it appears in the three areas of competence, although mainly in the area of Perspective.

As indicated in IPMA ICB4.0 [7], the strategy of the organization is the central place where aspects and principles of sustainability must be taken into account so that they influence the development of projects, programs or portfolios.

Regarding the rest of competence elements in the project domain, Otegi-Olaso [11] performs a detailed analysis of the direct references to sustainability that appear in the following competence elements: Strategy, Compliance, standards and regulations and Culture and values in the Perspective area; Personal integrity and reliability, Negotiation and Result orientation in the Practice area; and finally Scope and Change and transformation in the People area. In addition to these direct references, there are numerous references to aspects related to sustainability in the description of the different competence elements and their Key Indicators of Competence [12].

20.3 Methodology

This section presents the methodology used to achieve the objective of the research, which is a double objective. First, to investigate which aspects related to sustainability have been considered in the professional certification of project managers according to the IPMA standard ICB version 3. Second, to analyze how far these aspects are considered concerning the rest of the aspects taken into account.

The methodology comprises two steps: (1) identify the competence elements from the IPMA standard ICB version 3 that content sustainability aspects and (2) realize a data analysis of the competence elements evaluated in the certification process of level A (Certified Projects Director), B (Certified Senior Project Manager) and C (Certified Project Manager) assessed in the period between 2014 and 2019.

20.3.1 *Identification of Competence Elements Containing Sustainability Aspects*

The goal of this step is to identify the competence elements that contain aspects related to sustainability or sustainable development from the 46 competence elements that are part of the IPMA standard ICB version 3. In order to achieve this objective, the relevant literature on sustainability in project management is reviewed.

20.3.2 *Data Analysis*

In order to assess candidates, IPMA assessors select several competence elements from the 46 available according to ICB3. The number of selected CE varies depending on the level of the candidate and the development of the certification process. This certification process comprises several phases. The last two phases are the report and the interview.

Each level A, B and C candidates must submit a report that describes the application of the competences of project management in real cases. In ICB3 [6], there is a minimum number of elements of competence per area treated differently for each level, as shown in Table 20.2.

The report, together with other previous results such as self-assessment or written exam, will serve as the basis for the interview. As a background to each question in the interview, a real project management situation will be described in which at least one competence element with actions, tasks and results are involved. The interview must cover a minimum number of competence elements, as shown in Table 20.3.

The data analysis implies reviewing the work done by IPMA assessors, which allows the study to be carried out without accessing the reports of candidates, which

Table 20.2 Minimum number of CE per area treated in the report (reproduced from [6] with the permission of IPMA)

	IPMA level A	IPMA level B	IPMA level C
Technical	16	14	12
Behavioural	11	8	5
Contextual	8	6	4

Table 20.3 Minimum number of CE per area treated in the interview (reproduced from [6] with the permission of IPMA)

	IPMA level A	IPMA level B	IPMA level C
Technical	5–6	6–7	7–8
Behavioural	4–5	3–4	2–3
Contextual	4–5	3–4	2–3

are confidential. The tasks involved in the data analysis include extracting data from individual reports in level A, B and C certification processes, identify the competence elements evaluated in the STAR report, and the competence elements evaluated in the interview. As a result of this quantitative analysis, we should be able to obtain an ordered list of the competence elements according to the frequency that they have been selected by the assessors to evaluate the candidates. The elements of competence that, according to the literature, including aspects related to sustainability, are identified. In each case, the percentage of competence elements related to sustainability evaluated and the score obtained concerning the total are also studied.

Finally, in order to verify that the elements of competence mentioned in the literature are related to sustainability, a qualitative analysis is also carried out. The minutes of the interviews contain all the questions posed by IPMA assessors. From these minutes the questions raised for each of the elements of competence identified in the literature as related to sustainability are extracted. The revision of these questions allows to verify if the assessment of the elements of competence identified in the literature is related to sustainability. This work analyzes how the Spanish Certification Body assess, so the set of assessors has been considered as a whole, not individually. This avoids judging the performance of each assessor separately.

20.4 Results

This study has carried two different analyses: the first one, regarding the explicit relationship between competence and sustainability as a concept; the second one regarding the assessments of these competencies. In this section, results are shown and discussed.

20.4.1 Identification of Competence Elements Containing Sustainability Aspects

From the literature review, 11 competence elements are identified as having a relationship with sustainability or sustainable development concepts. This competence elements are 1.02 Interested parties and 1.04 Risk & opportunity from technical competences, 2.06 Openness, 2.09 Efficiency, 2.12 Conflict & crisis, 2.14 Values

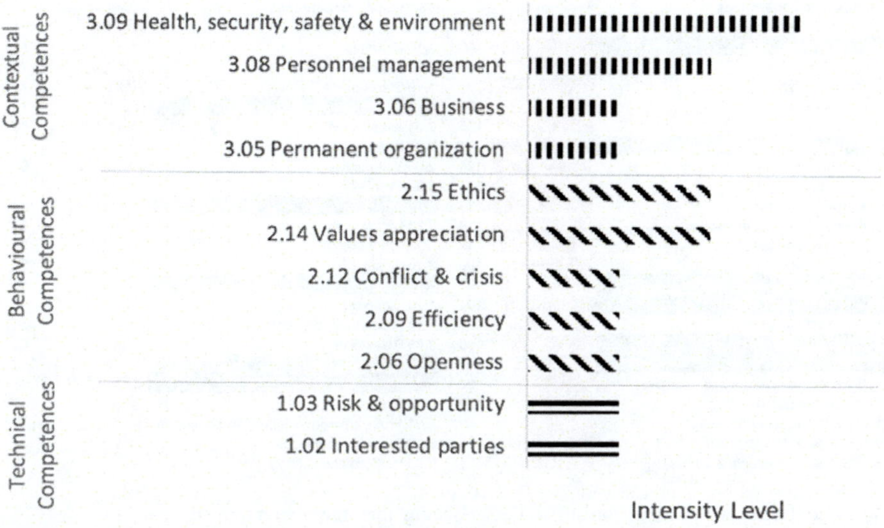

Fig. 20.1 Sustainability references in the standard IPMA ICB version 3

appreciation and 2.15 Ethics from behavioural competences, and finally 3.05 Permanent organization, 3.06 Business, 3.08 Personnel management and 3.09 Health, security, safety & environment from contextual competences. Figure 20.1 graphically shows the previous competence elements with an intensity scale that indicate the frequency with which various authors have identified them as related to sustainability.

As can be seen in Fig. 20.1, the element of competence "3.09 Health, security, safety and environment" is the one that has been identified most frequently.

20.4.2 Data Analysis

During the period from 2014 to June 2019 (month in which the last round of certification was held according to ICB3), the Spanish Certification Body certified and re-certified 497 candidates. Taking into account that level D and re-certified candidates do not need to elaborate a report, a total of 83 certifications (55 IPMA level C, 24 level B, and 4 level A) have been analyzed. The frequency analysis method was used to evaluate the data collected in the research. The frequencies of the appearance of the competence elements in the report and interview of the above certification processes show how many times IPMA assessors selected the element. In other words, it has been used to measure their significance in project management, according to the assessors.

Figure 20.2 shows the 46 ICB3 competence elements ordered according to their frequency of appearance in the certification processes carried out. As can be observed, of the first ten elements, 7 correspond to technical competences, 2 to behavioural

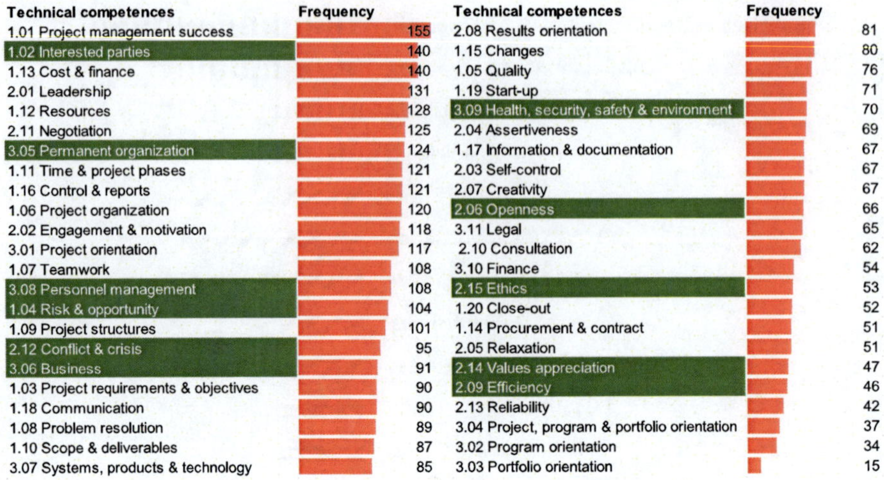

Fig. 20.2 Order of the CE according to the frequency they have been treated

competencies and 1 to contextual ones. On the other side of the last 10 elements, 2 correspond to technical competences, 5 to behavioural competencies and 3 to contextual ones. It should be noticed that technical competences have a higher weight than behavioural and contextual in the IPMA certification process, especially at level C, which in the case study represents 66.27% of the sample.

20.4.3 Results Discussion

From the literature review, it is observed that only two technical competence elements are related to sustainability: "Interested parties" and "Risk and opportunity". However, "Interested parties" is the second competence element only surpassed by "Project management success", and it is the most selected competence element among those related to sustainability. The competence element "Interested parties" is evaluated by assessors in 100% of the reports and in 69% of the interviews studied.

"Permanent organization" is the second more common element related to sustainability. As well as "Business", it includes references to sustainable aspects only in case the organization is adopting sustainability in its business strategy.

"Health, security, safety and environment" is one of the elements more directly related to environmental and social sustainability. It occupies the 7th position among sustainability-related elements, and the 28th position among the whole competence elements. Assessors evaluate this element in 48% of the reports, and 36% of the interviews studied.

Aspects with a high importance on social sustainability, such as "Ethics" and "Values appreciation", appear in the group of the 10 less essential elements. "Ethics" is evaluated in 27% of the reports, and 37% of the interviews studied.

Figure 20.3 shows the frequency of the appearance of competence elements related to sustainability in both the report and the interview. Several elements, such as "Interested parties", "Permanent organizations" or "Health, security, safety and environment", have a higher frequency in the written report, whereas those elements related to values, such as "Ethics", "Conflict and crisis" or "Values appreciation", have a higher frequency of appearance in the interview.

A qualitative analysis is carried out to verify to what extent the elements of competence mentioned in the literature are related to sustainability. The minutes of 83 interviews (55 IPMA level C, 24 level B and 4 level A) have been reviewed. The questions raised for each of the elements of competence identified in the literature as related to sustainability are checked. The revision of these questions allows to verify if the evaluation of the elements of competence identified in the literature is related to sustainability or not. And if it is related to sustainability, with which of its three pillars (economic, social or environmental) is related. The results of this qualitative analysis are shown in Fig. 20.4.

The study confirms that some of the elements of competence identified in the literature are related to sustainability in more than 80% of the questions posed. This is the case of "Conflict and crisis", "Values appreciation", "Ethics", "Personnel management" and "Health, security, safety and environment". But, on the opposite end, elements of competence such as "Risk and opportunity" or especially "Efficiency", are related to sustainability only in a few of questions posed.

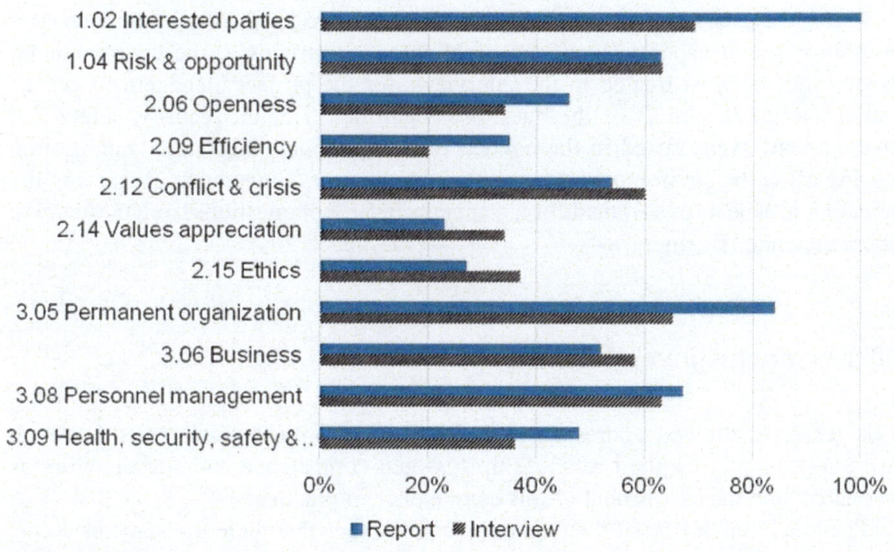

Fig. 20.3 Competence elements related to sustainability

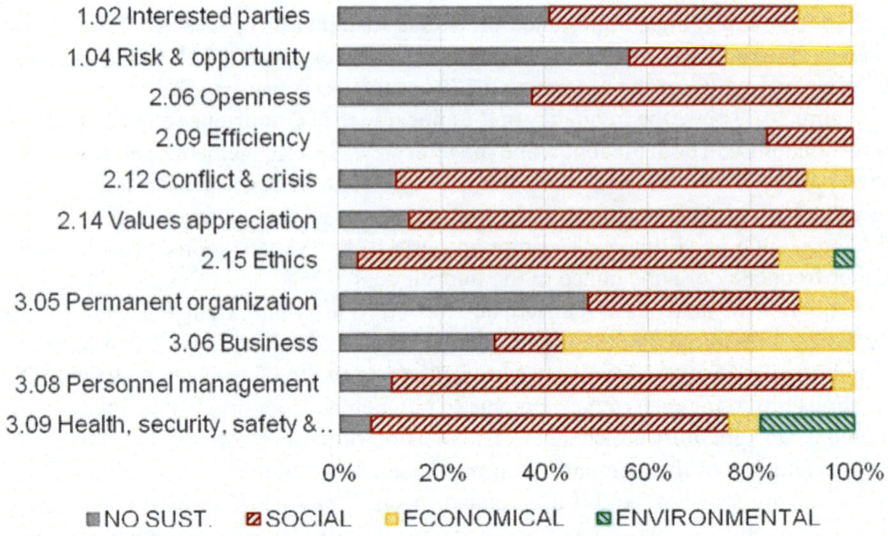

Fig. 20.4 Relationship between competence elements and sustainability, according to the questions of the interview conducted by the OCDP assessors

In the case of the element of competence "Efficiency", it must be said that, although conceptually it is closely linked to sustainability, it is one of the elements of competition that has been least treated in the evaluations studied (it is the fifth least frequently considered EC, see Fig. 20.2) and has only appeared in 20% of the interviews reviewed.

As for the aspects of sustainability that appear more frequently in the interview questions, social aspects should be highlighted. Surprisingly, the environmental aspects are the least treated in the interviews for the project management certification. Notice that most of the questions regarding "Health, security, safety and environment" were raised in the context of occupational safety and health, often leaving aside the environmental part of this element of competence. Regarding the economical aspects of sustainability, it appears in 57% of questions from the element of competence "Business".

20.5 Conclusion and Limitations

This research allowed addressing sustainability in project management through two questions: the explicit relationship between competence and sustainability as a concept, and the assessment of this competence in practice.

The main conclusion of the quantitative analysis is that there are some elements of competence more likely to be considered than others, but sustainability is not a factor that causes some elements of competence to be taken into account more than others.

Quantitative analysis of the data collected revealed that "Interested parties" is one of the most important elements of competence for the evaluation of competence. This EC is treated in 100% of the reports reviewed. "Health, security, safety and environment", one of the elements more directly related to environmental and social sustainability, occupies a middle position among sustainability-related elements, as well as among the whole competence elements. Aspects with a high importance on social sustainability, such as "Ethics" and "Values appreciation", appear in the group of the 10 less essential elements.

Qualitative analysis confirms that, in the practice, some of the elements of competence identified in the literature, such as "Conflict and crisis", "Values appreciation", "Ethics", "Personnel management" and "Health, security, safety and environment", are related to sustainability, whereas other elements of competence such as "Risk and opportunity" or especially "Efficiency", are related to sustainability only in a few of questions posed by assessors. Besides, this qualitative analysis shows that social aspects are treated with a high frequency in contrast with the environmental aspects of sustainability that are considered very rarely in the interviews for the project management certification, at least in relation to the elements of competence reviewed.

It was not possible to make a correlation between project typologies and competencies identified because that requires access to the candidates' files, which are confidential. The STAR reports of the candidates have not been accessed, but anonymous tables drawn from the minutes and the documentation generated by the evaluators have been used.

Another limitation of our study is that a correlation analysis has not been performed between the questions asked in the interview by each of the assessors to avoid judging the performance of the assessor. This work analyzes how the Spanish Certification Body assess, so the set of assessors has been treated as a whole, not individually.

The work carried out serves to prove the existence of a relationship between sustainability and competence assessment since sustainability is implicitly present in competence elements. In order to reach a greater detail and to be able to separate the relationship between competences and the three pillars of sustainability, it would be necessary to modify the assessment procedure.

Thanks to the explicit reference to sustainability in the new ICB4.0 (the word "Sustainability" appears in the three areas of competence, mainly in the area of Perspective), a future replication of this study applied to the ICB4.0 competence elements may demonstrate the relationship between sustainable and competent project managers.

We hope that this study and other similar ones carried out in other Certification Bodies will encourage this approach to be included in the training processes of assessors and that the CVMB (Certification and Validation Management Board) analyze the possibility of making the evaluation of sustainability even more explicit in future versions of the ICR.

Acknowledgements The authors wish to thank the board of the Spanish Certification Body for preparing and providing us with the anonymized version of the assessors' minutes in order to process them and carry out this work; and thanks also for the assessors that have authorized the access to their reports.

Compliance with Ethical Standards

All procedures performed in this study were in accordance with the ethical standards of the Certification Body of the Spanish IPMA's Member Association (OCDP). The authors, authorized by the OCDP (authorization with reference number OCDP QMS ISO17024-IPMA_ICB 20200420), conducted the study with an approach to knowledge and diagnosis of the situation of the sustainability within the IPMA certification process in Spain, with the aim of improving the OCDP performance.

Although it is a retrospective study, informed consent was obtained from all individual participants included in the study. The participant assessors were informed of the study scientific goals and accepted that their reports were used for this study.

The authors of the study have processed the data anonymously and have handled the information collected only for the purposes indicated. The candidates and assessors names were removed from the documents handled. Anonymous tables drawn from the minutes and the documentation generated by the assessors have been used. This work analyzes how the Spanish Certification Body assess, so the set of assessors has been treated as a whole, not individually. The information that was accessed was destroyed once the study was completed.

References

1. Aarseth W et al (2017) Project sustainability strategies: a systematic literature review, Int J Proj Manage 35(6):1071–1083. https://doi.org/10.1016/j.ijproman.2016.11.006 (Elsevier Ltd, APM, and IPMA)
2. AEIPRO (2017) AEIPRO—IPMA 4LC. Available at https://www.aeipro.com/es/certificacion/ipma-4lc. Accessed 30 May 2019
3. Buckle P, Thomas J (2003) Deconstructing project management: a gender analysis of project management guidelines. Int J Proj Manage 21:433–441
4. Eskerod P, Huemann M (2013) Sustainable development and project stakeholder management: what standards say. Int J Manag Proj Bus 6(1):36–50. https://doi.org/10.1108/17538371311291017
5. Gareis R, Huemann M, Martinuzzi A (2013) Project management and sustainable development principles. Project Management Institute, Inc., Newtown Square, PA
6. IPMA (2006) ICB International Competence Baseline. 3.0.
7. IPMA (2015) ICB International Competence Baseline. 4.0.
8. IPMA (2018) Certification—IPMA International Project Management Association. Available at https://www.ipma.world/individuals/certification/. Accessed 30 May 2019
9. Marcelino-Sádaba S, González-Jaen LF, Pérez-Ezcurdia A (2015) Using project management as a way to sustainability. From a comprehensive review to a framework definition. J Clean Prod 99:1–16. https://doi.org/10.1016/j.jclepro.2015.03.020 (Elsevier)
10. Økland A (2015) Gap analysis for incorporating sustainability in project management. Proc Comput Sci 64(1877):103–109. https://doi.org/10.1016/j.procs.2015.08.469 (Elsevier Masson SAS)
11. Otegi-Olaso JR (2016) Consideration of sustainability in project managers individual competence baseline. In: Proceedings from the 20th international congress on project management and engineering, Cartagena, July 2016, AEIPRO. http://dspace.aeipro.com/xmlui/handle/123456789/760

12. Paneque-de la Torre A, Bastante-Ceca MJ, Capuz-Rizo SF (2017) Análisis de los aspectos y principios relacionados con la sostenibilidad en la IPMA ICB4. In: Proceedings from the 21st international congress on project management and engineering (Cádiz, July 2017), AEIPRO. http://dspace.aeipro.com/xmlui/handle/123456789/336
13. Silvius AJG (2017) Sustainability as a new school of thought in project management. J Clean Prod 166:1479–1493. https://doi.org/10.1016/j.jclepro.2017.08.121
14. Silvius AJG et al (2017) Considering sustainability in project management decision making; an investigation using Q-methodology. Int J Proj Manag 35(6):1133–1150. https://doi.org/10.1016/j.ijproman.2017.01.011
15. Silvius AJG, van den Brink J, Kohler A (2010) The concept of sustainability and its application to project management. In Knoepfel H, Taylor T (eds) Proceedings of the international expert seminar in Zurich, Switzerland on 18th–19th Feb 2010. Zurich, Switzerland: IPMA, pp 131–145. 978-3-9523891-0-2

Chapter 21
Project Management Competences and the Sustainability of Non-profit Organisations

Amin Saidoun

Abstract The proposed paper addresses the need project management competences of non-for profit organisations in order to be sustainable in the longer run. The contribution of non-profit organisations (NPOs) to the economy is widely acknowledged among researchers. To manage projects in order to pursue their missions, NPOs need to remain sustainable. In this context, the paper addresses the importance of project management competences and their importance for the sustainability of NPOs, i.e. to exist and develop in the longer run. The empirical investigation undertaken in form of qualitative research offers an adequate approach to transition from literature review to some findings based on two case studies discussed in this paper. The first case concerns a medium-sized international fundamental research NPO based in Grenoble (France) that produces neutrons for science. The second case concerns a small national NPO which advocates the interests of large companies from the logistics sector based in Bonn (Germany). The discussion of the findings will conclude with suggestions for further research directions linked to the topic.

Keywords Competences · NPO · Project management · Sustainability

21.1 Introduction

A non-profit organization (NPO) works for any purpose other than making profits. At a more macro level of any economy and as underlined by the interviewed NPO leaders in cases presented in this paper, NPOs have an important role in society. They create value that is not always immediately measurable. This is what the two cases have in common. Unlike for profit sector where stakeholders look at quarterly reports and the profit/quick wins of the organization, the level of tolerance regarding the short-term performance and output of NPOs is much higher as they benefit society

A. Saidoun (✉)
Alma Mater Europaea, Maribor, Slovenia
e-mail: saidoun2015@gmail.com

Grenoble, France

R. Cuevas et al. (eds.), *Research on Project, Programme and Portfolio Management*, Lecture Notes in Management and Industrial Engineering,
https://doi.org/10.1007/978-3-030-60139-3_21

on the longer run. This concerns in particular fundamental research where tangible results will mostly be generated on the longer run.

A second argument for the importance of NPOs in society is that volunteers engaged in NPOs are essential to transmit values such as commitment and passion for work without necessarily getting any financial compensation. They work according to the principle: "I work/I share so I am", which is one of the major aspects that distinguishes NPOs from for profit organisations.

The expansion of non-profit sectors in most countries around the world over the past thirty years has generated increasing interest among researchers.

Only in India and in the United States to give an example, their number has doubled over the past decades [1].

However, the stopping of the activity of some NPOs raises the question of their sustainability, i.e. their ability to be sustainable in order to serve its stakeholders such as owners, clients, the community and the representatives of the profession in the longer run [11]. Deriving from this question, the importance project management competences for non-for profit organisations in view of keeping a NPO sustainable will be addressed.

In the first part, the paper will summarize some findings from the literature review on the importance of project management competences for NPOs.

In the second part, the paper will discuss the need of project management competences to sustain the NPO with two case studies elaborated by the author following an empirical investigation with NPO leaders. The first case concerns an NPO from the fundamental research area of an international institute based in France (Grenoble). The second case concerns a national NPO from the logistics sector based in Germany (Bonn).

Lastly, the author will come up with some recommendations for further research areas needed to better understand some findings derived from this empirical investigation.

21.2 Background and Objectives

While reviewing some literature on NPOs to produce this paper not many sources address individual project management competences of persons working in NPOs and how these competencies impact the sustainability of NPOs. These organizations provide goods and services that are not produced by the private sector, or at least not at the same price. They develop skills, provide jobs and help people to become social included [2, 9]. Over the last few decades, researchers insisted on the fact that NPOs must become more entrepreneurial in their processes and operations [15] and adopt innovative practices [11].

In this context, adopting strategies and to increasing the efficiency and effectiveness in project management have become critical for NPOs. In short, NPO leaders need to implement the ideas they have in mind.

According to some studies undertaken [3, 13] it appears that the sustainability of organisations has many facets that imply a change in which an organization adapts on different levels. On a first level, technical solutions need to be elaborated, corporate sustainability reports published and employees trained on sustainability basics (Case B Table 21.2). With this first level, the context to adopt sustainability practices is set [5]. The second level (also called value level), there is a need for changed employees' values and beliefs which include ethical and more responsible values [3]. The two levels of corporate sustainability can be seen in different perspectives regarding the culture of the organization [14]: the tip of iceberg, i.e. the part of culture that can be seen (the structure of organisations, the processes in place, the behaviours of persons, etc.), the values to which people adhere (strategies, goals and philosophies), and some assumptions (beliefs and perceptions which drive action).

After having reviewed recent literature on the sustainability of NPOs, the proposed paper addresses two research questions:

(1) How important are project management competences for NPOs?
(2) What are the main project management competences needed by non-for profit organisations in view of keeping them sustainable?

21.3 Research Methodology

The approach used to measure which competences matter most was a multiple case study as the adopted research method [10]. It was based on an interview including 27 questions carried out with NPO leaders in the fundamental research area and the logistics sector (Table 21.1). The choice of the case studies was done according to the following criteria: data access, data availability, experience of interviewed persons in the sectors and on the topic, duration of existence of the NPO. What differentiates both cases among other aspects are the size of the organization, the location, national versus international activity, the economic sector and the age. It was important to choose two cases and not only one because one NPO is successful its development and the second is declining in membership and less successful than in first NPOS to which the same questions were addressed. Following the classical multi-case qualitative research approach as described by Maylor et al. [10], the two cases described below will show common and points which will be identified and discussed.

Table 21.1 Characteristics of the cases

Case	Size of NPO	Person interviewed	Purpose	Focus/Location
A	Medium	Director General	Fundamental research with neutrons	International/France (Grenoble)
B	Small	Director	Logistics	National/Germany (Bonn)

Case A: *ILL* (www.ill.eu): "The ILL (The Institut Laue-Langevin) is an international research centre that dedicates its activity to neutron science and technology. It is known among researchers as the ILL produces neutrons with a very high flux reactor. Around 40 research instruments are used by scientists These instruments are set up near the beam lines that are coming out of the reactor. Every year, about 1400 researchers from over 40 countries visit the facility and 640 experiments selected by a scientific review committee are carried out. The fundamental research focuses on a variety of fields: condensed matter physics, chemistry, biology, nuclear physics and materials science, etc. While some researchers work on the design of engines, fuels, plastics and household products, others are looking at biological processes at nano-level (i.e. cellular and molecular level). The ILL has signed partnerships with R&D departments of industrial companies in order to perform fundamental science. ILL's scientists are also experts in neutron research and technology. This particular know-how supports the worldwide scientific community. With 500 full-time equivalent staff is mainly financed by France, Germany and the United Kingdom. 10 other countries support the financing at minority level. The NPO is undergoing a large programme to prepare for the future. It includes different projects and is called "Development plan 4.0". This programme as well as past programmes of the organization are divided into projects with a clear structure, project sponsors and regular project meetings, trainings and reportings to the steering committee. A project management handbook is also existing and updated according to the evolution of the project management culture which is much more established than in the organization of case B.

Case B: BvDP (www.bvdp.de): The BvDP (Bundesverband Deutscher Postdiensleister e.V.) is a German non-profit organization that carries out national and European the Dialogue with Parliaments, Governments and societal groups. It participates actively with position papers in the formation of public opinion. It analyses political developments and is responsible for market and sector intelligence. It initiates and supports scientific studies and expert opinions. It manages a large database on the postal and logistics sector. The BvDP brings together experts and coordinates and moderates the dialogue between corporate members. Through is member assembly, seminars and workshops and working groups the BvDP manages a network of high quality. It was founded in 1999 with a budget of 1,5 Mio Euro and is based in Bonn (Germany). Its members are large organisations such as Siemens, Deutsche Post/DHL, DHL express, Detecon Consulting, Diebold-Nixdorf, Lufthansa Industry Solutions, T-Systems, and more. It is financed by its members through annual membership fees and dedicated project funding. Operations and Projects are concentrated on the retaining of members, the acquisition of new members and fostering the relationship between the NPO and critical governmental institutions. The challenge of the NPO is that over the past years, members have left their organization, the objectives of the organization have changed and project management skills and trainings of top and middle management are missing.

The directors of the NPOs were given the questions before the interview to prepare for. The answers were collected from the interviewed persons during a face-to-face meeting, enabling the author to gather further data.

Some of the closed questions included a Likert scale to measure from 1 to 7 how important project management competences are for the NPO (1 low importance, 4 relevant, 7- = very important, see also attachment). By filtering the most important ones from the others, it is possible to get further information on the competences most needed in the NPOs.

21.4 Case Study and Interview Survey Findings

21.4.1 How Important Are Project Management Competences for NPO's Sustainability?

In addition to the operations and daily process work to make a NPO functioning, NPOs have projects to achieve their goals. These goals are derived from the visions, mission and strategy adopted by the owners of the NPO. The owners of an NPO can be governments like in the case of the Institut Max von Laue-Paul Langevin (ILL, case A), companies like for the Bunderverband Deutscher Postdiensleister (case B) or other NPOs (example: International Project Management Association which is a federation of national project management associations). They are usually represented by a steering committee (or council of delegates) that elects or commissions a board to manage the NPO and implement the NPO's projects. In the case A the trainings (both internal and external) provided in project management to the persons in charge in the organization on a regular basis are a proof of the awareness that project management are important to achieve NPO's goals. In the case B, trainings in project management is a luxury that is only satisfied by management on an exceptional basis. The fact that the executive director is longing for such trainings shows that the organization only builds on its own ability to manage projects based on common sense but without good practices as known and applied by the profession. What distinguishes both cases is that in the first case there are KPIs in place that are set for each strategic goal in place and can be seen in the development plan (Table 21.1). The case B no such KPIs are in place.

One could argue if we take the project management competence model of IPMA (2. ICB 4.0), if it would add value to introduce the concept of Key Competence Indicators. This would make sense if the NPO would adopt the ICB4 in its training ecosystem on project management which is still not the case.

When project management laid the foundation for the project management competence development (PMCD) framework introduced by Suikki and implemented in the company Nokia in 2001, the importance of project management competences in business (but not only) reached a new level of awareness [16].

As more organisations adopt project management approaches and the demand for project manages grows, competences of project managers become interesting for many stakeholders [4].

As an example, in case B, people from different horizons who do not know each other need to work together for the first time, often under time constraint. Here the NPO leaders clearly underlined the lack of PM competences that include tools and processes as element of practice competence. Examples that were quoted are: lack of project design, lack of project scoping and risk and opportunity management.

21.4.2 What Are the Most Important Project Management Competences Needed by NPOs to Sustain in the Longer Run?

21.4.2.1 Finance Competence

Finance comprises all activities necessary to estimate, plan, gain, spend and control financial resources both ingoing and outgoing to the project [7].

In both cases investigated NPO leaders ranked the need for finance as project management competence very high and also repeated. In case A three arguments were put forward: Firstly, the available resources of national budgets for fundamental research are limited. Only joining resources and in particular financial means enable to construct, run and maintain large-scale research facilities. Secondly, top researchers and talents are living all over the world. Thirdly, relying only on national researchers limits the potential for successful research, attracting funds and produce high-quality publications.

Two tangible examples can be quoted related to finance in project management competences:

– First example concerns funding/resources. One of the first reasons for uncertainty is government policies that change. Government policy changes have often. In the past, the organization was essentially relying on public funding coming from the research ministries of Germany, France and the UK. Since a couple of years, ILL (case A) is establishing partnerships with industrial companies interested in experiments that they could not perform on their own. For this purpose, private companies enter collaboration agreements with ILL in which teams of scientists and experts from the industry work together. They carry out experiments that have a direct impact on the industry. With this elegant type of collaborations, different stakeholders are satisfied. The ILL gets additional funding, private companies from the industry have access to a facility they could never afford on their own. The taxpayer gets a quick win out of the use of the research facilities as the experiments are less fundamental and more applied with quicker return on investments for the consumer and time to market. In short the competence element finance and the ability for a scientist to get grants and research funding has become a critical competence element for a scientist to do good science.

Government has been a cause of uncertainty as many initiatives did not receive the necessary political engagement [8]. "Reinventing government ideas [12] and new public management have transformed the relations between NPOs and the government. The practice confirms that “Endurance programme” mentioned by the interviewed director of the NPO/ILL is not guaranteed by the associates (government funding) for the year 2019 in terms of funding. This means either looking for alternative funding sources or to postpone the implementation of some parts of the programme.

In the case B competence in finance was also ranked very high, especially as the budget of the logistics federation has been shrinking since its creation in 1999 by one third. The reason for this is inherent to the activity of the federation that moved from a more public affairs focus to technology scouting in the logistics sector. The other reason is the lack of finance competence related to finding new models of engagement with companies from the logistics sector (or related to it).

Finance competence in project management means cost controls, engaging volunteers and searching for new members that are essentially corporate members. This is presently done with a membership discount scheme whereby if a member gets an attractive discount when it brings along a new member, in addition to the efforts deployed by the executive director (case B). This scheme is certainly good but it seems to be less attractive than in the past.

Finance competence also means searching for new funding opportunities through collaborations. There is a need to diversify the sources of income and reduces the dependence on project funded by the government. Some organizations focus on organization of special events, dedicated donor programmes and undertaking licencing initiatives [17]. At the ILL, the CRG scheme (collaborative research groups) can be seen as a kind of in kind contributions where scientists working on the payroll of international universities or labs are seconded at the ILL doing research. The output is shared in these collaborative agreements between ILL and the said labs to produce science for society.

21.4.2.2 Result Orientation

Result orientation is when an individual concentrates intensively on the outcomes of the project [7].

Large scale facilities work in an environment where as in the private sector, “coorpetition” (co-operate and compete) is not a buzz word. It forms part of every day reality. A tangible example for this is the five years time limited working contracts for scientists who need to prove their scientific output performance. The interview with the ILL director confirmed what the literature states on the need of non-profits to become business like in their operations. Here the slogan “if you want to be more charitable you need to be more business-like” finds its full meaning in the context of project management practice competences.

Vision statements made by the association of the NPO can only be implemented through projects that have sponsors and are split into clearly defined work packages. The characteristics of these work packages are spread among the divisions of the NPO with dedicated project managers.

Table 21.2 Key performance indicators at ILL

	Period	Target	Comment
Beam time delivery			
Number of cycles per year	2019–2028	4-4-3 after 2025	2 cycles in 2019 2 cycles in 2020 and 2 or 3 cycles in the period 2021–2025 adapted to the available financial resources
Number of operational instruments	2019–2028	28	+ 10 CRGs (Collaborative Research Group) Instruments

Source Development plan 4.0, Institut Max von Laue-Paul Langevin, Grenoble, 2019

The development plan 4.0 of ILL (case B) aims to achieve performance levels in line with the strategic envisaged goals. Each performance level is measured with an appropriate key performance indicator (see Table 21.2). The performance levels are ambitious but at the same time they are achievable with the available resources [6].

Result orientation as key success factor for the sustainability of the NPO means reducing the time to publication, the availability of software and expertise for data treatment. Result orientation means also that other divisions in the NPO such as the projects and technique division understand the need for good instrument maintenance to insure instrument performance and reliability. In case B, the result orientation was not necessarily seen as the most needed competence but evaluated as important.

21.4.2.3 Personal Communication Competence

Personal communication covers the sharing of adequate information, delivered precisely and consequently to all relevant parties [7].

An important project management competence mentioned in cases A and B was communication. By this, the NPO leaders meant both internal communication but also external communication which includes marketing and public relations. Questions such as "What are the outcomes we are getting out of this? What are the expectations of the people we are serving?" have become more than just a way to address the purpose of the NPO. Deriving from these questions adequate communication plans were elaborated in both cases (A and B).

An example to illustrate what precedes is the following: In the past the ILL had a motto/slogan: Neutrons for Sciences. This was recently changed to "ILL, Neutrons for Society". For different reasons, it took the organization almost 35 years to change this slogan. This change was enabled thanks to openness and self-reflection defined as key competence elements by the director of ILL who run successively different modernization and maintenance projects of the facility. By opening up to all society, i.e. by simply changing the slogan, the purpose of the organization or "raison d'être" was changed and much more understanding for neutron sciences addressing all parts of society was generated among society at large. In the case B, the communication of the strategic importance of the objective of the association was qualified as important. In practice, it was acknowledged that this competence remains a challenge.

21.4.2.4 Self-reflection and Self-management

Self-reflection is the capability to accept, reflect and analyse the emotions, behaviours, preferences and values that one has as well as their impact on our environment. Self-management is the ability to define goals, to control and adapt progress and to manage the work involved on a daily and systematic basis [7].

Self-reflection and self-management were ranked very high in terms of competence need. What confirms this need according to the director of the ILL is a fundamental change regarding the acceptance of project management.

This change in paradigm was accelerated by two main challenges the ILL had to overcome:

Firstly, security and risk management are topics that are inherent to the nature of a nuclear facility that produces neutrons with highly enriched uranium. Thus a high self-esteem, self-confidence and recognition of project management competences such as risk management are already existing for over 43 years at the ILL. Consequently, any input on the subject from outside the organization and the national security authority could/was considered in some cases as questionable. This attitude has tremendously changed after a worldwide known "Fukushima effect" and its impact on all nuclear facilities around the world, including the ILL. The result of that nuclear tragedy was that for the first time in history, a nuclear facility had to prove it could simultaneously resist to an earthquake of a certain magnitude and a mega flood, without jeopardizing the surrounding environment, especially the 600.000 inhabitants around the city of Grenoble where the ILL is based. External project management competences and advices on the subject were increasingly accepted.

The second challenge is the mind-set of some scientists regarding their role and status in the NPO. If one asked some scientists what it means to become a good scientist, the answer was simple: "I need to do good science". It was also underlined "I do not want to be downgraded to a portfolio, programme or project manager". This is why at some point in time the acceptance of project management training measures was rather low in some divisions of the ILL. Nowadays, after a shift in awareness about the importance of PM competences for the sustainability of the organization, the need and acceptance for more PM competences has increased at the ILL.

21.4.2.5 Resourcefulness

Rourcefulness is the capability to put in practice various techniques and ways of thinking to define, analyse, make priorities and find different ways to solve challenges and simulating solving challenges and problems. It often needs thinking and acting in an original and creative manner, and stimulating the creativity both at individual and team level. Resource fullness is useful when risks, opportunities, problems and challenging situations pop-up [7].

This is even more important in the context of a nuclear facility which has to follow special legal requirements in operations and management of projects. A strict control and assessment of both these two areas are imposed by the government in France. Any important derogation from these established rules needs prior government consent (consent of the French nuclear authority) otherwise the shut down of the reactor is imminent. The consequence of this would be the disruption of operations with highly negative international impact on the image of the ILL and in the scientific community. Scientific experiments are planned long time in advance by scientists coming to Grenoble from all over the world. The strict rules mentioned above, call for resourcefulness was ranked at 7 (very important) in the Likert scale in Case A.

21.4.2.6 Strategy Competence

Strategy competence underlines how strategies are understood, and implemented into tangible elements using projects [7].

Another competence that was highly ranked among the perspective competences in project management was strategy competence. Strategy development within the NPO requires the need to balance between different dimensions of the same task to realize a good integration and strategic focus between mission and money. The NPO leader mentioned, "I have to run a 100 Mio. Euro business with a clear social aim. My job is to make sure that the aim creates economic and scientific benefit to both the organization and also the broad society that the social outcome benefits both the staff of the ILL and the people we are serving. I have to be cautious in all I do and can't be too ambitious and remain strategic". In case B, strategy competence led to a change on the focus the work of the NPO, which was important to satisfy its members' needs.

21.4.2.7 Practice Competences

Regarding the practice competences as described in the standard ICB4 of IPMA, the following competences were highly ranked by the director: programme design, scoping, time, finance resources (ability to define, acquire and control and develop resources necessary for the project), plan and control, risk and opportunity and lastly stakeholder management. The director explained that for many years his organization was somehow cocooned from the world. He explained to which extent the context changed. This in turn calls for good practice competences at all level of the organization, which are changing over time.

21.5 Suggestions for Future Research

Having identified one of the limitations of this paper it would be interesting to undertake further research on Non-Governmental Organizations to widen the scope of the findings. To accomplish this further cases should be investigated and leaders from the NGOs (as an extension to NPOs) in international organisations should be interviewed accordingly. Further questions may be addressed in that context: How to measure the impact of project management competences on NPO's (NGOs) sustainability? What criteria should be chosen to measure the sustainability of NPOs?

21.6 Summary/Conclusion

The literature review and two case studies with leading project managers of a medium-sized international NPO from fundamental research and small national NPO from the logistics sector made clear: project management competences are needed by the organization to sustain in their respective competitive environment.

Firstly, an awareness of the importance on project competences is needed to achieve the NPOs objectives.

Secondly, the competences addressed were essentially individual competences of the top and middle-management of the organization.

Finance, result-orientation and effectiveness, self-reflection, culture and values and practice competences were identified in the case studies as most important ones in view of the sustainability of their organisations.

These individual competences are necessary to implement the work packages defined in an ambitious change/development programme to prepare for the horizon 2030 and beyond. The organization with less project management competence development and with less project management trainings (case B) is clearly less successful in the achievement of its objectives and mission that the organization of case A.

Project management competences will help top management to better explain to many stakeholders, including Steering Committees, employees and tax payers how well scarce resources are employed for the greater good of society in the longer run. Furthermore, and in a general context where NPO sector is becoming larger than ever and continues to grow, some studies reveal the percentage of contribution to the GDP of a country amounts to roughly 12%. It is therefore implicit that developing and implementing a project management culture in the NPO sector is a line to follow.

Lastly, the social mission of an NPO could be based on an economically viable organizational outset, which can be enabled by sound individual project management competences of the top and middle management.

Acknowledgements I would like to thank Ms. Constanta Bodea from University of Bucarest for her guidance for the qualitative research approach and structuring to the topic. I would like to thank Mr. Helmut Schober Director of Institut Max-von Laue—Paul Langevin in Grenoble (France) and Mr. Eugen Pink, Executive Director of Bundesverband Deutscher Postdienstleister e.V. in Bonn

(Germany) for the time dedicated for the interviews and the information provided to get an insight from the practice and share their experience as NPO leaders and project managers.

Compliance with Ethical Standards

I declare that informed consent was obtained from the participants, i.e. they knew what they were doing and they were agreeing with what they were doing. I confirm that the data was anonymised and that any special procedures to assure anonymity when data isn't anonymised have been made. According to the regulations of the Alma Mater Europaea University, this research does not require approval ethics from the ethics body of the university.

References

1. Casey J (2016) Comparing non profit sectors around the world. What do we know and how do we know it? J of Non profit Education and Leadership 6(3):187–223. http://dx.doi.org/10.18666/JNEL-2016-V6-I3-7583
2. CEEDR: Researching business support needs of ethnic minority owned business in Coventry and Warwickshire, Report to Coventry and Warwickshire Chamber of Commerce). Middlesex University, London (2001)
3. Crane A (2000) Corporate greening as amoralization. Organ Stud 21(4):673–696
4. Crawford L (2005) Senior management perceptions of project management competences. IJPM 23(1):7–16
5. Dunphy DC, Griffiths A, Benn S et al (2003) Organizational change for corporate sustainability: A guide for leaders and change agents of the future. Routledge, London
6. ILL development plan 4.0: Institut Max von Laue-Paul Langevin, Grenoble (2019)
7. IPMA Individual competence baseline for project, programme and portfolio management. International Project Management Association Available online at: http://products.ipma.world/wp-content/uploads/2016/03/IPMA_ICB_4_0_WEB.pdf, IPMA, Zürich (2015)
8. Kunle A (2004) Staffing, Retention, and government funding: a case study. Non profit Manag Leadership 14(4):453–465
9. Lyons M (2001) Third sector: the contribution of non-profit organisation and cooperative enterprises in Australia. Allen & Unwin
10. Maylor H, Blackmon K, Huemann M (2007) Researching business and management. Palgrove, London
11. McDonald RE (2007) An investigation of innovation in non profit organisations: the role of organizational mission. Non Profit Volunt Quart 36(2):256–281
12. Osborne DE, Gaebler T (1992) Reinventing government: How the entrepreneurial spirit is transforming the public sector. Addison-Wesley, Reading
13. Purser RE (1994) Guest editorial: "Shallow" versus "deep" organizational development and environmental sustainability. J Organ Change Manag 7(4):4–14
14. Schein EH (2004) Organizational culture and leadership, 3rd edn. Jossey-Bass, San Francisco
15. Sharir M, Lerner M (2006) Gauging success ventures initiated by individual social entrepreneurs. J World Bus 41(1):6–20
16. Suikki R, Tronsted R, Haapsalo H (2004) Project management competence development framework in turbulent business. Technovation 26(5–6):723–728
17. Weerawardena J, McDonald R, Mort GS (2010) Sustainability of non profit organisations: an empirical investigation. J World Bus 45(4):346–356

Index

R. Cuevas et al. (eds.), *Research on Project, Programme and Portfolio Management*, Lecture Notes in Management and Industrial Engineering,
https://doi.org/10.1007/978-3-030-60139-3

S

T

U

W